PROTEIN BIOTECHNOLOGY

By

Ashok Kumar
Lecturer
Department of Zoology
Bundelkhand University
Campus Department
Jhansi

DISCOVERY PUBLISHING HOUSE
NEW DELHI-110002

First Published-2006

ISBN 81-8356-133-0

Published by

DISCOVERY PUBLISHING HOUSE
4831/24, Ansari Road, Prahlad Street,
Darya Ganj, New Delhi-110002 (India)
Phone: 23279245 • Fax: 91-11-23253475
E-mail:dphtemp@indiatimes.com

Printed at:

Amit Enterprises

Preface

The present title **Protein Biotechnology** tries to capture the field of biotechnology in its whole some entirety in a manner that is concise and easily understood by the students. Enzymatic proteins play a vital role in degradation and fermentation providing a well known example but more recent developments of an over increasing range of raw materials have led to a wider and greater impact on the biotechnology industry.

Because it is undergoing such rapid growth, there will almost certainly be many examples of protein biotechnology in actual application when this book appears which had not reached that stage when it was written. For this reason we have allowed ourselves to include some examples which look as if they are likely to find application in the near future, despite our endeavour to illustrate the technology be examples that are actually in use, rather than possible processes that might one day find application but all too often really illustrate unfulfilled promise.

There can be no claim to originality except in the manner of treatment and much of the information has been obtained from the books and scientific journals available in different libraries.

The author expresses his thanks to his friends and colleagues whose constant inspiration have initiated him to bring out this book.

The author expresses his gratitute to Mr. Wasan and staff of M/s Discovery Publishing House for their whole hearted co-operation in the publication of this book.

Author

CONTENTS

1

Applied Protein Biotechnology: A Review

The major goal of protein engineering is the generation of novel molecules, intended as both proteins endowed with new functions by mutagenesis and completely novel molecules. This definition, which may sound broad and perhaps ambitious, in fact pinpoints one of the most promising developments in our ability to understand and control a protein's function. After the revolution introduced in protein science by the advent of genetic engineering, protein engineering can be considered as a second wave of innovation which is providing important breakthroughs in basic research and application, useful for studies on structure function relations and for exploitation in industry. Genetic engineering makes available unlimited amounts of purified proteins, whereas protein engineering produces tailor-made proteins redesigned such as to make them more suited to industrial requirements. On this basis it becomes evident that industrial biotechnology will enormously benefit of this possibility.

Natural targets of protein engineering are enzymes and several examples of modified catalysts have already been achieved and applied to industrial processes. However, protein engineering is not restricted to this field, since several non-enzyme proteins play important roles, for example, as drugs. Another fundamental area is that concerning antibodies, which may be planned both as specific carriers able to target drugs in the human body, as well as in the production of catalytic antibodies (enzymes) to be applied for reactions with non natural substrates.

Protein engineering is a complex and multidisciplinary field, where several different techniques and knowledge are applied in combination. The protein of interest needs to be first purified and characterized with regard to its functional properties, then to be cloned and over expressed in a suitable host organism and subsequently to be modified so as to improve its performances. A variety of analytical as well as structural techniques will be then employed for its characterization, whereas fermentation technologies will support its large-scale production. Different enabling technologies contribute therefore to an interdisciplinary approach that is usually represented as the protein engineering cycle.

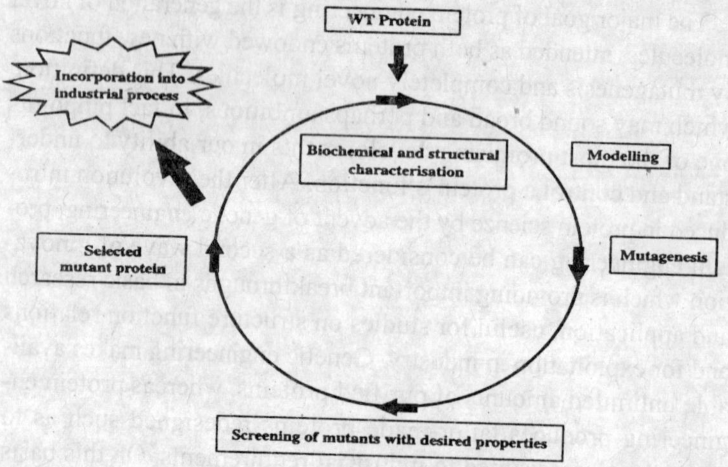

Figure 1.1 : *Cycle of protein engineering*

This chapter aims to outline the most innovative techniques as well as industrial applications of protein engineering, to provide the reader with a general, although necessarily non comprehensive, view on the field. More in general it attempts to "give the taste' of the opportunity provided by protein engineering to biotechnology, stressing the important and interactive linkage between basic science and application.

ENABLING TECHNOLOGIES

Diverse and complementary techniques form the knowledge and technological basis to engineer proteins and all of them must be mastered or at least be familiar to protein engineers. These basic techniques have been covered in recent and comprehensive books and will therefore not be recalled thereafter. However, new techniques of mutagenesis/screening were recently introduced, that expanded the possibility of modifying a protein's properties and deserve a brief comment. Other essential enabling technologies are also outlined in the following.

Techniques of Mutagenesis

The starting point for every step of mutagenesis is a DNA sequence cloned from the original source or synthesized based on the protein sequence of interest. One or several amino acids may be substituted/inserted/deleted using a wide variety of methods. Two main conceptual approaches can be followed: i) site-directed mutagenesis, i.e. substitutions of nucleotides in the correspondence of pre-selected positions in the gene (polypeptide) ii) random mutagenesis.

TABLE 1.1 : EXPERIMENTAL APPROACHES FOR PROTEIN ENGINEERING

Rational design

Allows for the introduction of mutations targeted to specific protein sites. Requires a detailed knowledge to the protein structure and of structure-function relationships.

Molecular evolution

Does not require any knowledge on the protein structure and mechanism of action. It is based on the random generation of a vast number of mutants followed by screening for the desired functions

Generation of random libraries

Production of large collections of proteins, peptides of region thereof. Is often coupled with surface display to ease screening of the mutants

De novo protein design

Generates novel structural scaffolds able to accommodate active sites or other protein functions

---

Rational design

Site directed mutagenesis (SDM) is a classical approach for protein engineers, exemplified by a broad scientific literature and by several contributions of this book. It consists in introducing a change in one or more amino acids and evaluating the effect of these pre-selected substitutions in the mutated product. By definition, this strategy, also called "rational mutagenesis", requires a prior knowledge of the role played by specific residues or regions of the protein. This means availability of the protein 3D structure, if possible also in complex with substrates. Ligands, regulation elements or at least availability of sequence of proteins with related but not identical activity for comparison. In fact, sequence alignments may also provide support in the selection of the positions to mutate, in particular when the protein of interest belongs to a large and well characterized family of proteins. On the other hand computer-assisted technology may help in predicting the functional effect of the planned substitutions.

Rational design is being very widely employed to engineer in proteins new functions or influence their regulation. Important goals have been achieved by rational design, as it will be briefly summarized in the following paragraphs.

Non-rational (random) design

Random mutagenesis on whole protein sequences or parts thereof is the method of choice in all those cases where knowledge about the structure and function of the protein of interest is not sufficient to support a rational design approach. The most innovative techniques involve the generation of repertoires (libraries) of mutated sequences and procedures of mutation/selection that mimic the processes followed by nature during evolution. In both cases, a very high number of mutant variants is produced, so that the development and availability of sensitive and fast procedures of screening is vital to manage the experimental work.

Molecular evolution

Very recently, protein engineering has been revolutionized by new methods that mimic evolution, of course in much shorter time, in that they allow for the accumulation and selection of mutations beneficial to the property of interest. These new approaches commonly referred to as *molecular evolution techniques*, allow

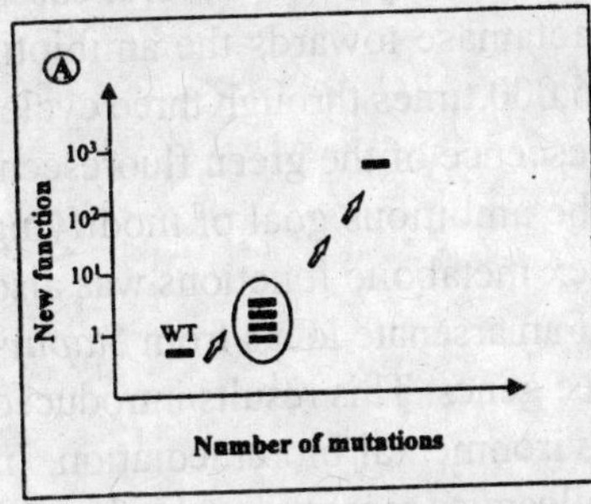

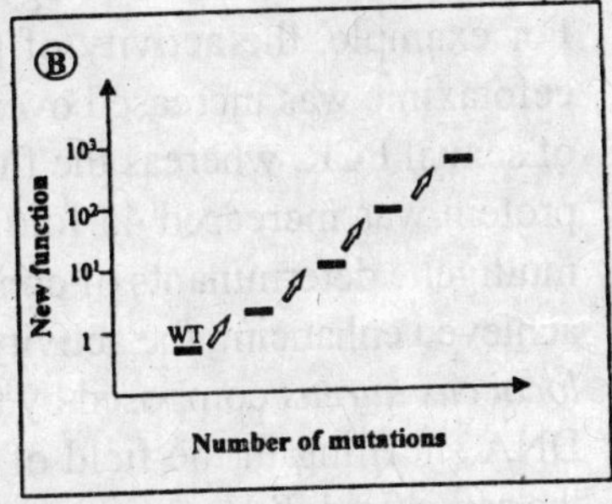

Figure 1.2 : *Comparison between two strategies for protein "evolution": DNA shuffling (A) and directed evolution (B)*

to overcome the serious drawback represented by the fact that for several industrially important proteins the level of available knowledge is insufficient to rational design. Moreover, growing evidence suggests that many protein functions cannot be ascribed to single or few amino acids but they rather depend on reactions located far away from the active and regulatory sites, and therefore hard to be predicted *a priori*. Molecular evolution methods are based on the accumulation of beneficial mutations over several rounds of mutagenesis. Since the effect of every single mutation can be small, the availability of methods of screening sensitive and in addition applicable to large number of mutants, is an indispensable prerequisite to this goal. Evolutionary approaches can be grouped in two categories: DNA shuffling and directed evolution. Both methods rely on the well-established technology of polymerase-chain reaction (PCR), in its development known as "error prone PCR", that allows for the introduction of a small preselected number of mutations in the DNA sequence.

DNA shuffling was developed by W. Stemmer in 1994. This technique is also known as "sexual PCR". It involves enzymatic fragmentation of DNA derived from different organisms previously randomly mutated as to introduce changes in the coding sequence. Fragmented DNA are then reassembled by errorprone PCR in the absence of primers, allowing their spontaneous reassemble and —at the same time—the introduction of new mutations. A final step of amplification with primers generates full-length products. This strategy has been applied to several cases. For example, the activity of β-lactamase towards the antibiotic cefotaxime was increased over 16,000 times through three cycles of sexual PCR, whereas the fluorescence of the green fluorescent protein was increased 45 fold. The ambitious goal of modifying multigene determinants of complex metabolic functions was also achieved enhancing the activity of an arsenate locus from *Staphylococcus aureus* composed by three genes. This results introduced DNA shuffling in the field of environmental bioremediation. In the same field, Kumamaru and colleagues recently evolved novel specificities in biphenyl dioxygenases (BPDox) by shuffling homologous genes from *Pseudomonas alcaligenes* and *P.cepacia;* mating yielded enzymes with enhanced 2,3-dioxygenation activity towards several biphenyl compounds and new oxygenation activity for aromatic hydrocarbons.

In vitro directed evolution is another powerful approach, developed by F. Arnold at Caltech, which with repetitive cycles of mutagenesis/screening forces the evolution of the selected sequence towards the desired functions. Differently from DNA shuffling, in this approach after every cycle of random mutagenesis, only a single variant showing functional improvements is selected and used as the parent for the next generation of mutants. A big advantage of this strategy is that it allows limiting the number of mutants to screen; on the other hand, by selecting a parent molecule at every step the researcher introduces a bias on the possible changes. This strategy has been successfully employed to engineer new functions as to make enzymes able to work in unusual environments. An esterase from *Bacillus subtilis* was modified as to become able to hydrolyze p-nitrobenzyl ester bonds in the presence of organic solvents, a reaction of interest for the pharmaceutical industry to remove protecting groups introduced during the synthesis of antibiotics of the class of cephaloporins. Specific activity was increased of 30-folds through four generations or random mutagenesis followed by DNA recombination. Even the specific activity of subtilisin E in polar solvent was enhanced through the same approach. Another property of paramount importance in biotechnology—enantioselectivity—was introduced by directed evolution in a lipase from *Pseudomonas aeruginosa*.

Combinatorial libraries and surface display

Another fundamental approach towards the generation of random mutants is the construction of "random libraries" representing peptides, cassette mutagenized proteins or antibodies, followed by the screening of variants with the desired functions. Random libraries might be obtained by i) random cassette mutagenesis or ii) by chemical synthesis. Whatever the origin, this generates very large repertoires of molecular diversity; which is evident if one considers that random mutagenesis on only three positions within a peptide will originate 8.000 (203) different combinations. The technology of random libraries has proved to be potentially very powerful depending on the ability to screen simultaneously a vast number of variants. Screening is usually based on the affinity of the new variants for the target molecules (antigens, enzyme substrates etc), a procedure which is generally referred as "panning". A support of paramount importance to the development and application of combinatorial libraries was provided by the advent of techniques allowing the display of variant peptides or epitopes at the surface of phages or various cell types. If the sequence to be panned is fused with that of genes coding for proteins of the virus coat, mutagenized peptides will be exposed at the virus surface, without impairing its replication ability. Billions of mutated peptides and proteins can be displayed at the surface of phage particles and be selected by affinity chromatography. The library is thus screened for biological function, i.e. for its binding affinity or enzymatic activity. One of the main application of combinatorial libraries concerns the improvement of antibodies binding abilities. In this approach, the variable regions of antibodies are amplified by polymerase chain reaction and then mutated as to develop binding affinity for new antigens. Application to human antibodies is intensively explored for diagnosis and therapy. However phages are not the only available tool for protein display. Thus for example, heterologous proteins have been targeted and anchored to the outer surface of yeast, mammalian and bacterial cells. Among bacteria, both Gram negative (*Salmonella, Escherichia coli*) and Gram positive (*Staphylococci, Streptococci*) have been used, where gene fragments can be inserted into the genes coding for outer membrane proteins, lipoproteins, fimbria or flagellar proteins or whatever make fusion products accessible on the outer surface. Enzymes such as pullulanase and cellulase have been displayed on *E. coli* and staphylococci. Peculiar applications of surface technology can be found in the production of whole cells bioabsorbents for environmental purposes.

Application of the combinatorial approach to the identification of molecules with a particular biological activity is reported in the chapter by Tramontanto and Felici using a series of case studies ranging form the development of molecules that mimic carbohydrate antigenes and therefore useful for vaccines, to epitope mapping to the so-called "minibody", a small protein designed as to display antigen binding loops that can be randomised and used to select ligands.

Structure Analysis and Prediction

Structures and modelling

Besides developments in molecular biology and biochemistry, biocomputing and methods for the determination of 3D structures, are indispensable tools in this field. Since the three-dimensional structure of a protein (enzyme, hormone, receptor) determines its biological function, it becomes important to have available a high resolution 3D structure, especially when the protein engineer aims to redesigning the protein function. Therefore, the importance of structural information for protein engineering cannot be overestimated, since the immediate challenge is to understand the rules relating the primary sequence to its 3D structure and to determine how structure is related to function. Recent advances in protein structure determination techniques originated the number of new structure solved every year to exponentially increase. A big number and a big deal of information; however only a small amount with respect to the variety of forms and shapes available in nature.

The most classical approach to protein structure is still X-ray diffraction, through which resolutions at less than 1Å have been obtained. Important advances were made in the field by using more intense X-ray sources, as cyclotron facilities, two dimensional area detectors coupled to computers making faster the obtainments and analysis of structural data. Rapid acquisition times provide the possibility to study transient phenomena. Difficulties intrinsic to this technique, in particular the necessity to obtain suitable crystals of the proteins to be studied, has been recently relieved by the use of robots to perform crystallization. It is now in use to perform protein engineering to facilitate protein crystallization, for example by introducing cysteines to obtain heavy atom derivative crystals isomorphous to the native protein.

A very comprehensive example about the role of crystallography in the elucidation of structure-function relationships in enzymes is provided in the chapter by Longhi and colleagues dealing with cutinase, a small hydrolytic enzyme presently under study for applications in biocatalysis and detergents.

More recently, advances in multidimensional NMR (Nuclear Magnetic Resonance) methodology made this technique competitive, especially because of the possibility to study proteins in solution, eventually under conditions near to the physiological ones. It is presently possible to determine by NMR structures of proteins in the range of 15–35 kDa Mr, some of which in complexes with ligands, what is of importance for the study of protein-drug and protein-protein complexes.

The role of biocomputing

Nature provided us with an astonishing variety of shapes and functions, most of which have still not be unraveled by researchers.Thus, even if a few hundreds of protein structures are solved, some other thousands still await to be studied. Fortunately, evolution seems to have selected a limited number of protein folds, where different sequences can be accommodated. This fact allows in many instances to extrapolate from the known structure information about those we still do not know. Therefore, knowledge-based structural prediction is becoming a common approach. Protein sequences are aligned with the sequences of homologous peptides of known structure and a structural fram work is "modeld" by assembling places of structures derived from the most homologous proteins. This approach relies on the availability of structural databases and of software for sequence analysis and energy minimization as well as of computer graphic tools. One of the most innovative approaches in biocomputing is that known as *fold recognition*, a computational attempting to fish out the fold of a protein even when it has very little sequence homology to other sequences of the same structure.

Moreover development of computational methods in combination with computer graphics, presently accessible to most laboratories, introduced a significant element of rationality in protein engineering allowing to predict with an acceptable degree of confidence the effects of aminoacid substitutions. However, structure and sequence analysis still remains one of the bottlenecks in protein science.

TABLE 1.2 : BOTTLENECKS IN THE FIELD OF PROTEIN ENGINEERING

- Scarce information on the second genetic code, i.e. rules relating sequence to structure in proteins
- Limited number of known 3D structures
- Limited or not obvious information of the structure-function relations in proteins
- Difficulties in the management of information

The impact of biocomputing on biotechnology is getting increasingly dramatic due to the enormous amount of information continuously provided by the genome sequencing projects. Currently some dozens of projects of sequencing are underway, covering prokaryotic organisms unicellular eukaryotes and complex eukaryotic genomes). Enormous support is provided to researchers by the availability to the public of databanks and tools for analysis.

TABLE 1.3 : WHOLE GENOMES SEQUENCING PROJECTS (ARCHAE, BACTERIA, EUKARYOTE, NOT VIRUSES)

20 Complete Prokaryote Genomes
1 Complete Eukaryote Genome
45 Prokaryote Genomes in Progress
17 Eukaryote Genomes in Progress

De novo design

Differently from every approach mentioned before, where existing polypeptide templates are modified such as to introduce new properties, the aim of *de novo* protein design is to be able to design new (not pre-existing) sequences endowed with predefined structure and function. In this case the bottleneck is not a technical one, since synthesis of peptides can be now realised in several

TABLE 1.4 : SOME WEB SITES OF INTEREST TO PROTEIN ENGINEERS

Structures and structure analysis	
Protein Data Bank, Brookhaven National Lab.	http://www.pdb.bnl.gov
SCOP-Structural classification of proteins	http://scop.mrc-lmb.cam.ac.uk/scop
RasMol	http://www.umass.edu/microbio/rasmol
Sequences and sequence analysis	
European Bioinformatic Institute, UK	http//www2.ebi.ac.uk
European Molecular Biology Lab., G	http://www.embl-heidelberg.de
Expasy Molecular Biology Server	http://www.expasy.ch
NCBI Genebank	http://www.ncbi.nlm.nih.gov
Pedro's Biomolecular Research Tools	http://www.public.iastate.edu.

governing protein folding — what is often referred to as the second genetic code — are still not completely understood. This makes it difficult bot to predict protein structures from sequences and to design sequences that fold into a desired structure. The major difficulty arises from the enormous number of possible sequences to consider in the computational approach. As a result, designed structures have often a molten globule-like structure since they are not able to perform the tightly packed organization typical of native proteins. Recent developments obtained in automated protein design programs greatly extended the researcher's ability but still the complete design of novel biocatalysts exceeds the present know-how. However, small proteins able to adopt defined secondary structures have been designed and the number of sequence candidates to consider can be reduced by reducing the length of the sequence to be designed or introducing a bias in the search of experimental sequences, for example by phage display. One interesting achievement in this field was the construction of a hyperthermophilic variant of the B1 domain of Streptococcal proteins G. Among the first practical applications of *de novo* design one might recall the design of a four helix bundle protein used as a source of rare amino acids when expressed in the rumen bacteria of dairy cows and the design of albebetin and interferon fragment was attached and the recombinant protein proved to effectively bind to the murine thymocyte receptor activating the blast transformation.

Medium engineering and bioimprinting

It is worth mentioning that other approaches besides protein engineering, have recently improved our ability to control protein conformation and enzyme activity, i.e. medium engineering and molecular bioimprinting. Although these two approaches do not directly influence a protein's primary sequence, they produce conformational effects affecting its function, and are therefore properly considered in this book. The fields of medium engineering developed based on the observation that the performances and even the specificity of enzymatic proteins can be modulated through the solvent where the reaction is made to occur. Enzymology in near anhydrous environment, i.e. in organic solvents, is an important area in biotransformation. Three major benefits are related to the use of organic solvents: i) the higher solubility of several organic substrates; ii) the higher stability of the enzymes, for example to the exposure to high temperatures required by some processes; iii) an increased selectivity observed in several cases and concerning in particular selectivity towards enantiomeres compounds. The two latter properties have been suggested to depend on a certain degree of rigidity of the protein caused by the organic solvent itself. In some cases, the advantages obtained largely compensated for the decrease in activity produced by the anhydrous solvent. Often enzymes exposed to particular pH or ligands before liophylization and subsequent dissolution in organic solvents, prove to be able to retain specific properties, i.e affinity for the ligand, a property referred to as "memory". Based on this observation, a strategy for the activation of proteins has been developed which was called "molecular (bio)imprinting" and can be adapted to particular conditions where the reaction of interest has to take place. Supercritical fluids have also be exploited as non conventional reaction media. The contribution by Cernia and colleagues in this volume effectively illustrates how the specificity and activity of microbial lipases can be modulated through the use of organic solvents and supercritical fluids.

TARGET MOLECULES

Engineering Enzymes

The study of structure/function relations unnaturally occurring enzymes can provide important clues of interest for protein engineers. However, the variety of strategies employed by nature in developing enzymes adapted for specific functions and environments is still far to be completely exploited. On the contrary, it is reasonable to suppose that we just scraped the surface of the diversity of solutions evolved. Biocatalysts purified, cloned and employed in industrial processes are possibly only those which were possible to screen from easily available sources that, in addition should have produced the desired enzyme in sufficient quantity.

Novel screening approaches are revealing the existence of a variety of unsuspected enzymatic activities. Even if the natural proteins might be not directly useful to industrial processes, they can provide precious informations useful to redesign conventional enzymes to perform special (novel) activities.

In this frame, we should recall innovative screening procedures set up to allow isolation of enzymes which escaped previous investigation. One of this methods is the so-called "expression cloning strategy", that is based on the expression of cDNA libraries from the selected source organism in host. In this approach, transformation are screened for several different activities, thus allowing to isolate new enzymes without relaying on previous knowledge on the enzyme itself, because screening is completely activity-based. In this way Novo Nordisk was able to isolate over 130 fungal genes coding for potential industrial enzymes. In the same way, accumulation of more and more information about sequences of related enzyme, allows to develop nearly universal primers to be used in molecular screening programs. Recently, the ability to screen genomic DNA without the necessity to prior isolate the source organism has been reported, what obviously further extend our ability to obtain novel enzymes.

The search of enzymes from extremophiles is also passing through a real boom. Environments extremes as for high or low temperatures, pressure, salt concentrations are systematically exploited through highly concerted programs. The first commercial thermophilic enzyme has in fact been isolated over 30 year ago in thermal pools of the Yellowstone National Park. It is the DNA polymerase from *Thermus aquaticus* (Taq) which is now so widely used in PCR experiments. Extreme enzymes are seeked for use in processing of foods and chemicals or in some cases as leads for new antibiotics. The most important screening programs have been DEEPSTAR, that allowed to collect 1000 different organisms from deep-sea mud and 2500 strains, from the Marianas Trench and a European Project in the frame of the EC-Biotech program. Interestingly, strains were isolated able to grow on toxic organic solvents, with interesting perspectives for use an environment bioremediation.

Much is expected from the analysis of sequence and structure of extreme enzymes, since it may provide clues in understanding factors influencing in particular the stability of proteins in particular solvents.

Properties of importance to engineer are stability (to temperature, pH, specific reaction conditions), specificity, regulation and eventually properties useful for the downstream processing of the recombinant proteins.

Engineering stability

Stability is a feature of critical importance in biotechnology and it is therefore pursued with intense efforts. Stability may concern: stability to temperature (food and chemical processing), to pH (detergents), resistance to proteolytic cleavage (drugs). Attempts to understand factors affecting stability started quite early in protein science and have been based mainly on mutagenesis approaches or on the comparison between homologous sequences from, for example, mesophiles and extremophiles organisms.

Presently, several protein properties are recognised as having a stabilizing effect but we are still far from the possibility to control protein stability. Often, attempts to stabilized proteins by protein engineering have failed, since the introduction of stabilizing interactions destroyed other important interactions or caused a loss in function.

Several interactions of importance have been determined, each of these which give contributions which are quantitatively similar; the most significant are briefly outlined in the following.

- Disulfide bridge stabilize proteins by restricting the possible conformations of the unfolded proteins. However, engineering disulfide bridges has met alternate success, possibly because of the difficulty in predicting how the increased rigidity of the protein will be detrimental for the protein function. A significant success has been described in the stabilization of the T4 lysozyme by the Matthews group.
- Ionic interactions or salt bridges. Several thermophilic proteins have additional ion pairs as compared to their mesophilic homologous; however the effect of this interactions can be hardly forseen.
- α-helices stabilization is another common approach. Multiple mutations in the triosephosphate isomerase dimer have been introduced as to increase the helix propensity of some

protein regions or to stabilize the 2D structure itself by interaction with the helix macrodipole. Stability was increased of about 3.0 Kcal/mol thus approaching the stability of the same protein from the thermophilic bacterium *Bacillus stearothermophilus*.

- Hydrophobic effect. Is the well known effect by which hydrophobic regions tend to be sequestered in the protein interior. Even the tight and highly organized packing of the protein core is important and in fact the native state of proteins packing of the protein core is important and in fact the native state of proteins show packing densities similar to those of organic crystals. Substitutions of large hydrophobic residues in the protein cores with smaller amino acids (large to small substitutions, LTS) have destabilizing effect, probably because cavities are created within the protein structure. The opposite substitution (small to large, STL) and even double and compatible LTS and STL mutations have been considered as a possible way to improve stability. Recently, by this approach Shih *et al.* (1995) mutated three buried residues of chicken lysozyme obtaining to increase thermostability of about 0.6–1.3 Kcal/mol. Relying on the same effect but using random mutagenesis, Lee and colleagues (1996) were able to stabilize α1 -antitrypsin of 1–3.6 Kcal/mol.

Binding of ion metals or cofactors can increase structural stability when they bind with higher affinity to the folded than to the unfolded state. Engineering metal centers, in particular binding sites for metal ions, frequently Ca^{2+}, Zn^{2+}, Cu^{2+} aims to enhance protein stability but also to control activity or ease purification. The same concept of differently stabilized conformational states allows for control of protein activity; following this approach, histidine was engineered in the active site of trypsin to create a metal dependent switch of enzyme activity. Metal binding sites are also extremely useful to construct proteins that can be easily purified from complex media by interaction with immobilized chelating groups. Artificial metalloproteins are also explored as biosensor to detect the presence of free metals in solution. Moreover, modification of the activity of existing metalloenzyme is under study for use in bioremediation; in this case, the introduction of new binding pockets allows transformation of non-natural substrates, as it was recently obtain for monooxygenases modified such as to oxidize aryl sulfide and styrene. The most recent progresses in this field have been recently reviewed.

Not only stability to temperature is important in industrial processes. The lipase from *Candida antarctica* B has been mutated, to increase the oxidation stability of the enzyme, whereas always a lipase for detergents was modified as to increase its stability to alkali and to proteases as required to improve its washing performances.

Stability in organic solvents or other unusual reaction media that might be necessary during industrial processes. The ability to use proteins in unusual or non-natural environments greatly expands their potential applications in biotechnology.

Approaches to improve the stability of industrial enzymes when exposed to non-physiological conditions are reported in the chapter by Grand *et al.*, describing how Bacterial neutral proteases, isoamylases and carbamylases have been subjected to either site-directed or random mutagenesis to improve their performances in industrial processes.

Engineering specificity

Application of protein engineering to the specificity of enzymes and other proteins concerns such a vast field of action, making it difficult to be summarized. It might be directed towards the fine tuning of enzyme specificity, as well as the introduction of novel properties on preexisting protein scaffolds and the introduction of regulatory sites or switches. Rational design, random mutagenesis as well as molecular evolution techniques, have all been applied in the effort of controlling protein specificity, an issue that we will try to focus on through a few significant example.

A common approach towards redesigning enzyme specificity is that of using existing protein scaffolds to introduce new properties. Recently, a single amino acid substitution was sufficient to convert the protease papain to peptide nitrile hydratase.

A possible alternative is to construct hybrid proteins where different catalytic and/or regulatory modules are incorporated.

Binding sites for metal ions often prove useful to regulate enzyme activity. Thus the introduction of a binding site for Cu^{2+} was sufficient to allow regulation of trypsin activity.

Cofactor requirement can also be altered by mutagenesis. As an example we can quote the paper by Dean and coworkers that changed the dependence of *E.coli* isocitrate dehydrogenase from NADP to NAD by replacing amino acids in the adenosine-binding pocket.

Several studies successfully obtained changes in the enzyme substrate specificity. Alignment of sequences of homologous enzymes with different specificities as well as structural information have proven useful as a guide to identify residues possibly involved in this property. Thus for example serine proteases have been subjected to extensive mutagenesis studies aimed to covert one protease in another endowed with different specificity (i.e. chymotrypsin and trypsin). Serine proteases have conserved folds and sequences but differ in the specificity towards the point were they cut the peptide, being large hydrophobic residues for chymotrypsin and a positively charged residue for trypsin. The story of engineering serine proteases has been very informative. In fact, mutation of all four different residues in the S1 site (binding for the residue proximal to the cutting site) was not sufficient to convert an enzyme in the other. Otherwise trypsin acquired chymotrypsin specificity after replacement of two surface loops which are not obviously involved neither in substrate binding nor in catalysis. Following these studies, a good degree of control on the proteases specificity has been gained. Subtilisin for example was much studied. More easily than in the previous studies, the specificity of the *B. amyloliquefaciens* subtilisin was changed so as to be similar to that of *B. licheniformis* subtilisin by changing only three residues and the ability to selectively cleave two consecutive basic residues was engineered by Ballinger and colleagues

(1996) in the same enzyme.

Olesen and *Breddam* in their contribution to this book report about the production of mutants of yeast carboxypeptidase with altered P1 preference, with application for C-terminal modification of peptides and stereo-specific synthesis of peptides and peptide derivatives.

Less widespread is the approach to transfer enzyme active sites to stucturally unrelated proteins since this implies a heavy work of structural modeling.

Engineering of protein purification

An important issue for the production of industrial recombinant proteins is the possibility to obtain them in a pure and active form. Two contribution of this volume specifically focus on the introduction of "tags" allowing purification and folding of the protein of interest. *Hober* and *Uhlen* describe the use of fusion proteins and co-expression strategies to ease folding and recovery of recombinant proteins. *Schmidt* and *Skerra* developed by molecular repertoire techniques a novel tag able to bind with high affinity to streptavidin from where it can be eluted by competition with biotin analogues.

Engineering Antibodies

Antibodies are important targets for protein engineering that are opening a new era for their application in health care, agriculture and industry. The two major techniques in this category concern re-engineering of natural antibodies and catalytic antibodies. A very remarkable progress in the field was generated by the application of novel techniques, in particular of combinatorial antibody libraries displayed on the surface of phages. This technique, developed in the early 1990s, revolutionises the generation of monoclonal antibodies and their engineering, since the use of animals or hybridomas cultures is not longer necessary, being recombinant antibodies expressed in bacteria, yeast, plants. Antibody phage display is obtained by fusing the sequences coding for the variable region of immunoglobulins to proteins of phage coat. The expressed fusion protein is incorporated in the envelope of the mature phage particle being presented at its surface. This accessibility is the basis for selection procedures consisting, for example, in binding with immobilized antigenes and the enrichment of phage antibodies specific for the desired antigen. In this way large repertoires of antibodies can be created and subjected to subsequent screening. The coding sequence is in addition amenable to any further modification by mutagenesis and protein engineering techniques. One of the main applications of this strategy is the generation of antibodies with high affinity for antigenes that do not necessarily need to have been previously purified, as it was the case of the previously used immunization procedures. On the contrary, antibody's diversity is first randomly generated and specificity is obtained through the subsequent screening against even complex materials containing the antigen of interest. Researcher's ability to control the antibody specificity has increased so to make possible the rational design of tailor-made antibodies against any antigen of choice. In the field of health care, the possibility of producing recombinant human antibodies ev. engineered to increase their specificity, open completely new perspectives as for drug targeting. Immunotoxins, immunoconjugates and bispecific antibodies can be generated for application in therapy and diagnosis. The present state of antibody engineering for health care has been recently reviewed.

In the field of agriculture, engineered antibodies are studied as substitutes of plant hormones or to decrease the need of pesticides. Finally, the use of engineered antibodies during the removal of organic pollutants have been described.

In the second approach, the ability of the immune system to generate an enormous diversity of binding sites is exploited as a possible source of new catalysts. The concept underlying the generation of antibodies-based catalyst—catalytic antibodies or enzymes—is that, if an antibody is created with an antigen binding site able to bind to analogues of the transition states of chemical reactions, binding will lower the transition energy of the reaction. Through immunization with transition state analogues, monoclonal antibodies have been obtained able to catalyze a wide range of reactions, eventually not performed by natural enzymes. Catalytic efficiency however, is still lower than with enzymes. Newest developments in abzymes technology exploit combinatorial libraries displayed on the surface of phages, a methods which enormously enlarge the possible size of the library. Abzymes have been used to activate prodrugs.

Engineering signal molecules

Peptide hormones, receptors and in general component of the cell signaling machinery are all important targets for protein engineering, in view of novel applications in the field of health care. It was demonstrated that hormone/receptor interactions mostly depend on interactions among secondary structure elements. This observation led to the design of polypeptides with reduced size but enhanced biological activity, where secondary structure homology rather than sequence homology was maintained.

PRESENT APPLICATIONS OF PROTEIN ENGINEERING

Protein engineering is a pervasive technology and is a powerful tool to overcome instrinsic limitations weakening industrial applications of available proteins. The properties to introduce in a commercial protein strongly depend on the process where this molecule has to be employed and rather can be a combination of multiple *ad hoc* features.

Industrial Enzymology

Enzymes to be employed for industrial transformations have been the first targets of protein engineering and several engineered proteases, amylases, lipases, cellulases and xylanases have been introduced in the chemical, food and, more recently, environmental protection industry. The market for industrial enzymes in fact is steadily growing and is expected to further increase as scientists will become able to generate novel catalysts with optimized performances for the transformation process.

Most desirable features to be introduced being that molecules are stable and active under unusual conditions of temperature, pH or are active in unusual environments, as for example, in organic solvents. Besides, one should recall another property that may be of importance in industrial processes, i.e. the surface properties of the used catalysers. In fact, several important industrial enzymes, for example cellulases, xylanases and lipases rely for their activity on their ability to bind or adsorb to substrates and interfaces.

Hereafter a brief overview of protein engineering performed on industrial enzymes is presented, where both rational design and directed evolution have been successfully applied.

Proteases are the main class of industrial catalyzers, particularly because of their use in detergents. They account for over 40% of the global enzyme market. Proteases are added to detergents because of their action on protein stain (spots). Subtilisin in particular has been the target of in-depth engineering: its stability in the chelating environment typical of detergents has been increased of about 1.000 folds by deleting a calcium binding site, whereas the introduction of surface negative charges improved its laundry performances. More in general serine proteases have been one preferential target for protein engineers, to modulate their specificity and/or regulation.

TABLE 1.5: INDUSTRIAL ENZYMES TARGETS OF PROTEIN ENGINEERING

Starch processing enzymes: α-amylase, glucoamylase, glucose isomerase
Detergent enzymes: proteases, lipases amylases
Texiles: cellulases, amilases, lipases, catalases, peroxidases
Leader industry: proteases, lipases
Pulp and paper: xylanases, cellulases

Several instances of engineered enzymes for food processing have been reported. Among them, *glucose isomerase* used in the conversion of corn starch to high fructose syrup for use in sweeteners, in particular for soft drinks. One of the most desirable technical improvements would be a higher tolerance to temperature. Rational approaches have been applied to stabilize the interactions helding together enzyme monomers in the quaternary structure. *Calf chymosin* has been subjected to multiple point mutations to made specificity pH-dependent thus allowing modulation of flavor development. *Papain* has been modified to make it active even at low pH whereas phospholipase A2 termal stability has been increased through mutations in α-helical regions. Amylases are employed to convert starch in simpler sugars. First α-amylase yields maltodextrins which are then processed by glucoamylase giving glucose and β-amylase resulting in maltose. Application of processing is carried out at high temperature and there several protein engineering studies have been aimed at improving the thermal stability of the enzymes involved. Protein engineering of glucoamylase from *Aspergillus niger* is the topic of the contribution by *Frandsen et al.* that, based on an extensive mutagenesis work, provides insight in the mechanism of action, stability and specificity of this enzyme.

Cellulases and *xylanases* are used in the processing of textiles and in detergents. Most commercial cellulases are endoglucanases and contain a catalytic region and a cellulose-binding domain. Efforts have been attempted to increase the binding ability of these enzymes to their substrates as well as enzyme activity. In this book, β-glycoside hydrolases are the topic of the chapter by R.A.J. Warren, focusing on the basis of their thermal stability, pH optimum, catalytic activity and specificity, as elucidated by protein engineering. These enzymes hydrolyse chitin, cellulose, xylan and mannan as well as oligosaccharides used in the detergent, food processing and paper industry.

Pectinolytic enzymes find application in both food and pharmaceutical industry. The state of the knowledge about structure of this enzymes, regions involved in substrate recognition, mechanism of action and structural basis of specificity are reporter by *Mayans* and *colleagues* in their contribution.

Lipases are employed in several sectors, food, chemical transformations and detergents. Several mutagenesis studies have been applied to improve their performances. Fungal lipases used in detergents have been modified as to improve their resistance to proteases also present in cleaning agents and to change surface properties and make them more soluble; in addition several agents present in detergents such as surfactants and calcium sequestering agents reduce lipase activity and have been therefore targets for protein engineering. Site directed mutagenesis has been applied to the control of selectivity for chain length specificity towards triglycerides, towards triradyglycerols, whereas directed evolution methods yielded an increased enantioselectivity. Lipases are the subject of some contributions in this book. *Lotti* and *Alberghina* report on studies on the lipases from *Candida rugosa*, underlying the importance of evolutionary analysis for protein engineering of this family of isoenzymes. The paper by Bornscheuer *et al*. illustrates the state of the art on the lipases produced by *Rhizopus species*, enclosing X-ray crystallography, computer modelling and site-directed mutagenesis, opening wide application in lipid modification and organic synthesis. Finally, *Canaan* and *colleagues* describe the development of gastric acidic lipases for possible applications in substitutive enzyme therapy for exocrine pancreatic insufficiency.

Protein Engineering for Health Care

Only a few engineered proteins are already on the market for health care products but perspectives in this field are so promising that an explosion of application is easily to foresee, following the completion of clinical tests. The efficacy of a drug if dependent not only on its activity and selectivity but also on stability, specific targeting and rate of clearance in the body, all issue that are being addressed through protein engineering.

Peptides and *peptidomimetics* are probes that mimic the molecular interactions of natural proteins and are therefore useful in the process of drug discovery.

Peptide and *polypeptide vaccines*, especially for protection against hazardous viruses, since they mimic immunogenic epitopes of the viral coat. Hybrid proteins are constructed also with the purpose of correctly presenting the antigen to the immune system to elicit immune response. Hybrid vaccines already on the market against malaria and hepatitis. The chapter by *Grandi et al*. illustrates the role of protein engineering to design vaccine components with enhanced immunogenicity and devoided of any toxicity as well as for mucosal delivery of vaccines.

A big deal of work is devoted to the immune system, especially to the production of *engineered antibodies* for medical therapy and diagnosis: 30% of all proteins under clinical trials are antibodies.

Component of the intracellular *signal transduction machinery* and proteins involved in the regulation of gene transcription. Recently, superactive analogues for human glycoprotein hormones are being developed with potential applications in the therapy of thyroid and reproductive disorders. Knowledge-based mutagenesis was performed on a protein domain rich in basic residues involved in receptor binding and signal transduction, leading to the development of human thyroid stimulating hormone and chorionic gonadotropin analogues with increased receptor binding affinity and bioactivity. Injection of a variant of the IL-6 receptor proved able to induce a strong anti hIL-6 antibody response; the elicited antibodies bound circulating IL-6 with high affinity thus masking its biological activity, a result very promising for the treatment of immune and neoplastic diseases. *Vanoni* and *colleagues* in their contribution propose an approach to disease therapy based on the inhibition of a specific pathway of signal transduction, acting on a ras-specific guanine nucleotide exchange factor.

Protein engineering of *hormone receptors* aims to develop controllable gene expression systems regulated through the administration of small drugs, an experimental system allowing to dissect the mechanism of action. These systems require the expression of genetically engineered chimeric transcription factors that function as molecular or gene switches and regulate the expression of an exogenous target gene in response to administration of a drug. Gene switches are based on the genetic engineering of chimeric intracellular receptors that contain functional domains including a ligand binding domain, a DNA binding domain and a transactivation (transrepression) domain. Gene switches have been developed based on steroid hormone receptors. Estrogen receptors and their inhibitors as targets towards the discovery of new drugs are the topics of the chapter by *Connor* and *McDonnell*.

Koller and *colleagues* deal with surface receptors, suggesting protein engineering techniques to produce soluble receptors to be used in high throughput screens.

Growth factors are also considered as promising targets for mutagenesis, especially towards the development of wound repair drugs. Several companies are being involved in the search of non-peptide analogs of growth factors, i.e. fibroblast growth factor and platelet-derived growth factor.

Generally speaking, *bioscreening*, i.e. methods used to evaluate the molecular interactions of candidate drugs with their targets, is a central issue in drug discovery. Techniques of bioscreening in oncology are outlined in the paper by *Draetta* and *Boisclair*, with focus on their central role in the process of drug discovery.

Of course several *enzymes* of importance as drugs are possible targets for modifications. Among them we would like to mention recent studies on serine proteases, in particular thrombin. This enzyme is subjected to Na^+-dependent allosteric regulation: suppressing Na^+ binding by mutagenesis shift the balance between the procoagulant and anticoagulant activities of the enzyme, with perspective of application in the field of anticoagulant drugs.

Finally, *drug delivery* will strongly benefit of the use of fusion proteins designed as to selectively target disease-causing cells and extracellular targets; recombinant fusion proteins are under clinical trial for the treatment of diseases including tumor types.

Enzymes for Diagnosis and Rehabilitation

In the field of bioremediation the major contribution of protein engineering concerns the development of enzymes or modified microbial species able to degrade xenobiotic compounds released in the environment by human activities. We might recall the two previously mentioned exampled where molecular evolution methods allowed for the production enzymes degradating polychlorinated biphenyls and recombinant bacteria endowed with the ability to detoxify soluble arsenate released during mining. This book reports two contribution focussing on environmental issues. The development of biosensors, in particular based on fusion proteins useful for enzyme immunoassay and bioluminescent enzyme immunoassay to detect environmentally hazardous compounds is the topic of the paper by *Aizawa* and *colleagues*. Enzymes involved in the degradation of chlorinated compounds, one major class of recalcitrant polluting compounds are related in-depth by the contribution of *Janssen* and *colleagues*.

2

Amino Acids

Amino acids are the building blocks of **proteins.** Proteins of all species, from bacteria to humans, are made up from the same set of **20 standard amino acids.** Nineteen of these are α–amino acids with a **primary amino group** $\left(-NH_3^+\right)$ and a **carboxylic acid** (carboxyl; –COOH) group attached to a central carbon atom, which is called the α –carbon atom (C_α) atom is a hydrogen atom and a **variable side-chain** or 'R' group. The one exception to this general structure is proline, which has a secondary amino group and is really an α **-imino acid.** The names of the amino acids are often abbreviated, either to three letters or to a single letter. Thus, for example, proline is abbreviated to Pro or P.

STEREOISOMERS

All of the amino acids, except for glycine (Gly or G;), have four different groups arranged **tetrahedrally** around the central C_α atom, and thus can exist in one of two **stereoisomers.** These two stereoisomers (or **enant-iomers**) are **nonsuperimposable, mirror images** and are termed the D and L forms. Stereoisomers are physically and chemically indistinguishable by most techniques, but can be distinguished on the basis of their different optical rotation of plane polarized light. Molecules are classified as dextrorotatory (D; Greek 'dextro' = right) or levorotatory (L;

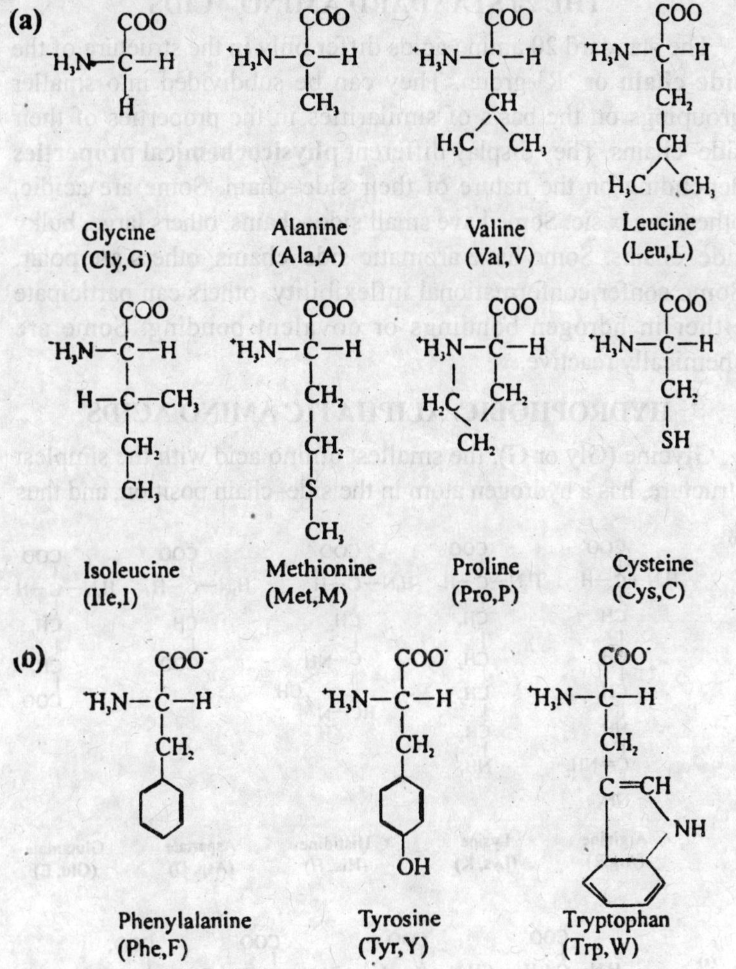

(a)

Glycine (Cly,G) | Alanine (Ala,A) | Valine (Val,V) | Leucine (Leu,L)

Isoleucine (Ile,I) | Methionine (Met,M) | Proline (Pro,P) | Cysteine (Cys,C)

(b)

Phenylalanine (Phe,F) | Tyrosine (Tyr,Y) | Tryptophan (Trp,W)

Figure. 2.1: ***The 20 standard amino acids,***

(a) Hydrophobic, aliphatic R groups, (b) hydrophobic, aromatic R groups.

Greek *levo*' = left) depending on whether they rotate the plane of plane–polarized light clockwise or anticlockwise. Only the **L–amino acids** are found in proteins **D–Amino acids** rarely occur in nature, but are found in bacterial cell walls and certain antibiotics.

THE 20 STANDARD AMINO ACIDS

The standard 20 amino acids differ only in the structure of the **side-chain** or 'R' group. They can be subdivided into smaller groupings on the basis of similarities in the properties of their side–chains. They display different **physicochemical properties** depending on the nature of their side–chain. Some are acidic, others are basic. Some have small side–chains, others large, bulky side–chains. Some have aromatic side–chains, others are polar. Some confer conformational inflexibility. others can participate either in hdrogen bondings or covalent bonding. Some are chemically reactive.

HYDROPHOBIC, ALIPHATIC AMINO ACIDS

Glycine (Gly or G), the smallest amino acid with the simplest structure, has a hydrogen atom in the side–chain position, and thus

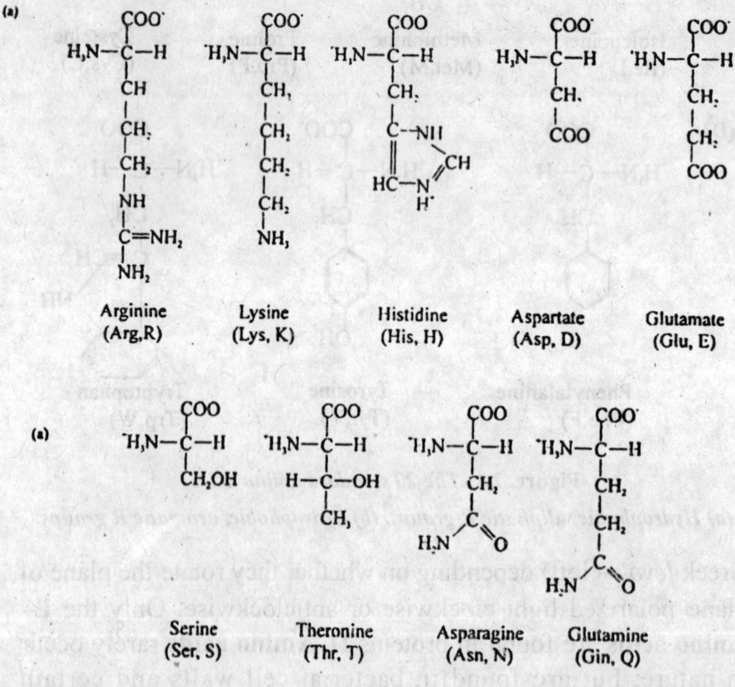

(a)

Arginine (Arg,R) | Lysine (Lys, K) | Histidine (His, H) | Aspartate (Asp, D) | Glutamate (Glu, E)

(a)

Serine (Ser, S) | Theronine (Thr, T) | Asparagine (Asn, N) | Glutamine (Gin, Q)

Figure. 2.2: ***The 20 standard amino acids (a) Polar, charged R group), (b) Polar, uncharged R groups.***

does not exist as a pair of stereoisomers since there are two–identical groups (hydrogen atoms) attached to the C_α atom. The aliphatic side–chains of **alanine** (Ala or A), **valine** (Val or G), **leucine** (Leu or L), **isoleucine** (lle or l) and **methionine** (Met or M) are chemically unreactive, but hydrophobic in nature. **Proline** (Pro or P) is also hydrophobic but, with its aliphatic side– chain bonded back on to the amino group, it is conformationally rigid. The sulfur–containing side–chain of **cysteine** (Cys or C) is also hydrophobic and is highly reactive, capable of reacting with another cysteine to form a disulfide bond.

Hydrophobic, aromatic amino acids

Phenylalanine (Phe or F), tyrosine (Tyr or Y) and **tryptophan** (Tyr or W) are hydrophobic by virtue of their aromatic rings.

Polar, charged amino acids

The remaining amino acids all have polar, hydrophilic side-chains, some of which are changed at neutral pH. The amino groups on the **side-chains** of the basic amino acids **arginine** (Arg or R) and **lysine** (Lys or K) are protonated and thus positively charged at neutral pH. The side-chain of **histidine** (His or H) can be either positively charge or uncharged at neutral pH. In contrast, at neutral pH the carboxyl groups on the side-chains of the acidic amino acids **aspartic acid** (aspartate: asp or D) and **glutamic acid** (glutamate; Glu or E) are de-protonated and possess a negative charge.

Polar, uncharged amino acids

The side chains of **asparagine** (Asn or N) and **glutamine** (Gln or Q) the amide derivatives of Asp and Glu, respectively, are uncharged but can participate in hydrogen bonding. **Serine** (Ser or S) and **threonine** (Thr of T) are polar amino acids due to the reactive hydroxyl group in the side-chain, and can also participate in hydrogen bonding (as can the hydroxyl group of the aromatic amino acid Tyr).

CLASSIFICATION

Amino acids can be classified in two ways based on structure and polarity.

Based on structure, amino acids are grouped into three classes as:

1. Aliphatic amino acids: These are straight or open chain amino acids which are further subdivided into four groups as:

1. **Monoaminomonocarboxylic (neutral) amino acids:** These consist of one amino and one carboxyl groups and hence are neutral to litmus, e.g. glycine, valine, leucine. iso-leucine, serine and threonine.
2. **Moonaminodicarboxylic (acidic) amino acids:** Thesé consist of one amino and two carboxyl groups and hence are acidic to litmus, e.g. aspartic acid and glutamic acid.
3. **Monocarboxylicdiamino (basic) acids:** These consist of one carboxyl and two amino groups and hence are basic to litmus, e.g. lysine, arginine and histidine.
4. **Sulphur-containing amino acids:** These consist of one or more sulphur atoms, e.g. cysteine, cystine and methionine.

II. Aromatic amino acids: These contain an aromatic ring in the molecule, e.g. phenylalanine and tyrosine.

III. Heterocyclic amino acids: These contain a heterocyclic nucleus in the molecule, e.g. histidine, tryptophan, proline and hydroxyproline.

Structure and range of amino acids

In an amino acid there is a central carbon atom, known as the α carbon atoms, to which is always attached an acidic carboxyl group, **–COOH,** a basic amino group, **–NH_2** and a **hydrogen** atom. The forth position is the only variable part of the molecule. The group there is known as the **R group**. This group gives each amino acid its uniqueness. The simplest amino acid, and therefore the **easiest one** to learn as an example, is **glycine,** where R is simply hydrogen. When R is –CH_3, the amino acid **alanine** is formed.

Rare amino acids

A small number of rare amino acids occur in the proteins of organisms. They are made from some of the common amino acids.

For example, hydroxyproline is made from proline, and is found in the protein collagen; hydroxylysine is made from lysine, and is found in collagen.

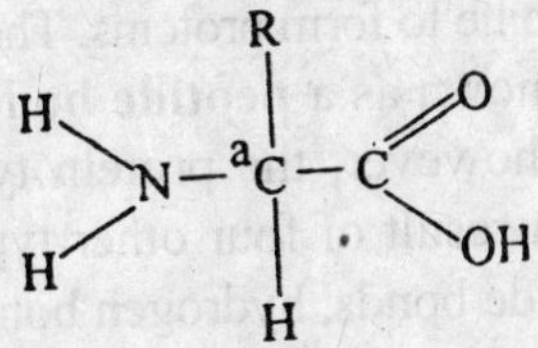

Figure. 2.3: ***General formula of an amino acid.***

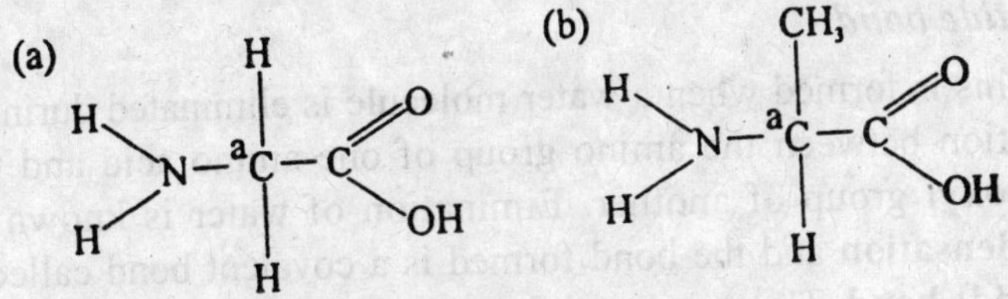

Figure. 2.4: ***(a) Glycine. (b) Alanine.***

There is no DNA code for the rare amino acids, and they are made from their parent amino acids after they have been incorporated into a protein.

Non–protein amino acids

Over 150 of these are known to occur, either free or in a combined form in cells, but never in proteins. For example GABA γ aminobutyric acid) is virtually unique to the nervous system. It is an inhibitory neurotransmitter, important in the brain.

AMINO ACIDS ARE AMPHOTERIC

Molecules like amino acids which contain both acid and a basic part are described as **amphoteric.** They are mainly as ions and can carry both a positive charge on the basic part and a negative charge on the acid part. Such ions are described as dipolar and are called zwitterions. This explains the fact that amino acids and proteins can be made to move in an electrical field, as when they are separated by electrophoresis. The charge on the amino acid can be affected by changes in it environment. For example, making a solution more acid would increase the concentration of hydrogen ions, and hence positive charge, and these would tend to cancel out the negative charges on the amino acid.

BONDS USED IN PROTEIN STRUCTURE

Amino acids combine to form proteins. They are joined together by a type of bond known as a **peptide bond.** Once these bonds have been formed, however, the protein typically folds into a particular shape as a result of four other types of bond, namely ionic bonds, disulphide bonds, hydrogen bonds, and hydrophobic interactions. Studying these bonds us to understand the structure and behaviour of proteins.

Peptide bond

This is formed when a water molecule is eliminated during a reaction between the amino group of one amino acid and the carboxyl group of another. Eimination of water is known as **condensation** and the bond formed is a covalent bond called a **peptide bond.** The compound formed is a **dipeptide.** It possesses a free amino group at one end, and a free carboxyl group at the other. This enables further combination between the dipeptide and other amino acids. If many amino acids are joined together in this way, a **polypeptide** is formed.

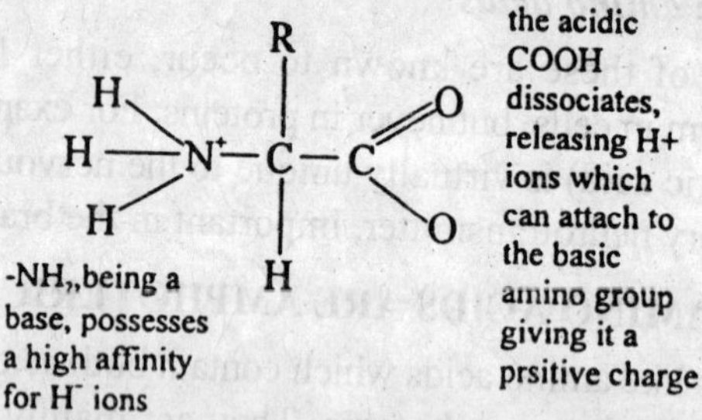

Figure. 2.5: ***Neutral zwitterion form of an amino acid.***

Acidic and basic R groups exist in an ionised (charged) state at certain pHs. Acidic R groups are negatively charged and basic R groups are positively charged. They can therefore be attracted to each other, forming ionic bonds. In an aqueous environment this bond is much weaker than a covalent bond and can be broken by changing the pH of the medium. This helps to explain the

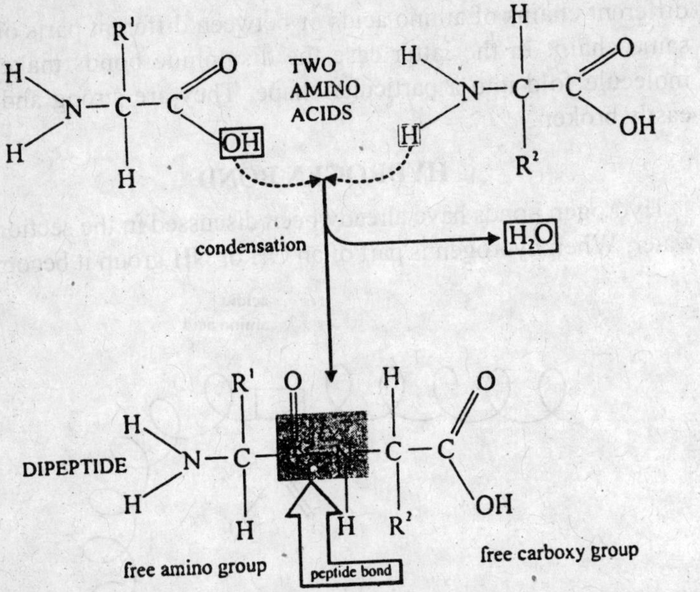

Figure.2.6: *Formation of a dipeptide by a condensation reaction between two amino acids. Condensation is removal of water.*

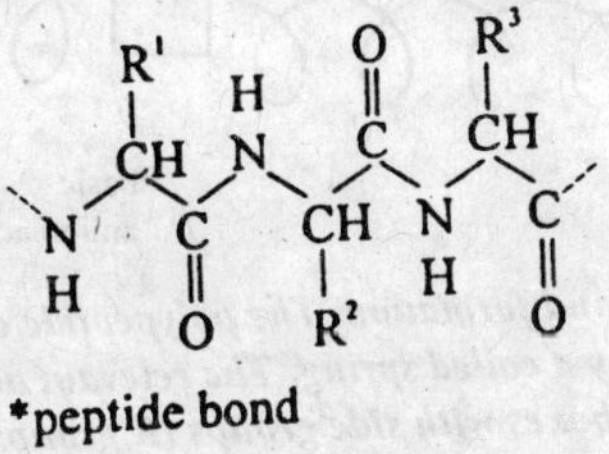

Figure. 2.7: *Part of a polypeptide showing the joining of three amino acids.*

desruptive effect that changes in pH can protein structure. For example, adding acid to milk makes it curdle because the ionic bonds in casein (milk protein) are broken and the protein ceases to be soluble.

DISULPHIDE BOND

The amino acid cysteine contains a sulphydryl group. –SH, in its R groups. If two molecules of cysteine line up alongside each other, neighbouring sulphydryl groups can be oxidized and form a disulphide bond. Disulphide bonds may be formed between different chains of amino acids or between different parts of the same chain. In the latter case the disulphide bonds make the molecule fold into a particular shape. They are strong and not easily broken.

HYDROGEN BOND

Hydrogen bonds have already been discussed in the section on water. When hydrogen is part of an OH or NH group it becomes

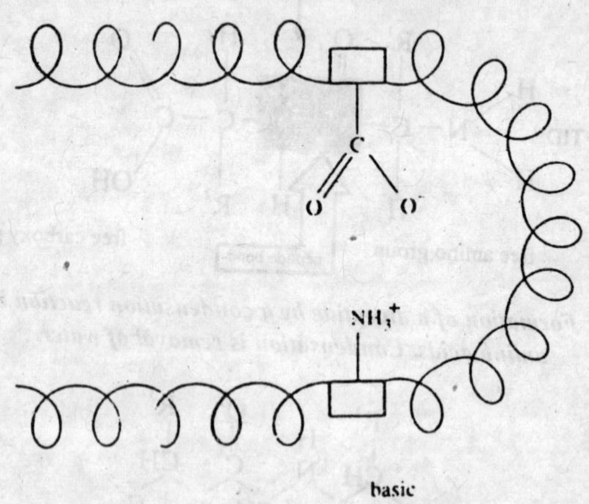

Figure. 2.8: ***Ionic bond formation. The polypeptide chain of the protein is represented by a coiled spring. The relevant amino acids are shown as boxes with side groups (R groups) attached.***

slightly positively charged (electropositive). This is because the electrons that are shaped, and which are negatively charged, are attracted more towards the O or N atoms. The hydrogen may then be attracted towards a neighbouring electronegative oxygen or nitrogen atom, such as the O of a C=O group or the N of an NH group C=O and NH groups occur along the length of polypeptide chains, and they can interact to produce regular shapes such as the α -helix discussed later. The hydrogen bond is weak, but as its occurrence is frequent, the total effect makes a considerable contribution towards molecular stability, as in the structure of the α -helix and silk.

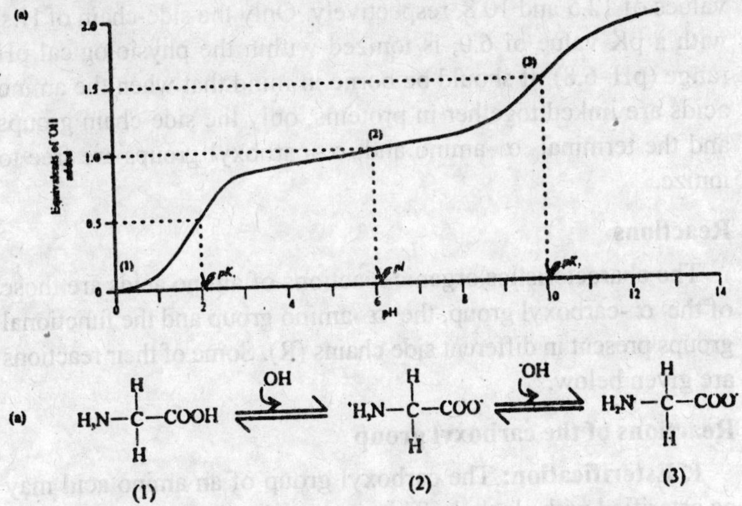

Figure. 2.9: ***Ionization of glycine (a) Titration curve of glycine, (b) dissociation of glycine.***

IONIZATION

The 20 standard amino acids have two **acid–base groups;** the α **-amino** and –carboxyl groups attached to the C_{α} atom. Those amino acids with an **ionizable side–chain** (Asp, Glu, Arg, Lys, His, Cys, Tyr) have an additional acid-base group. The **titration curve** of Gly is shown in. At low ph (i.e. high hydrogen ion concentration) both the amino group and the carboxyl group are fully protonated so that the amino group and the carboxyl group are fully protonated so that the amino acid is in the cationic form H_3N^+ CH_2COOH. As the amino acid in solution is titrated with increasing amounts of a strong base (e.g. NaOH), it loses two protons first from the carboxyl group which has the lower pK value (pK = 2.3) (the pK at which this group is half dissociated) and then from the amino group which has the higher pK value (pK = 9.6). The pH at which Gly has no net charge is termed its **isoelectric point, pI.** The α - carboxyl groups of all the 20 standard amino acids Asp and Glu have pK values in the range 8.8–10.8. the side-chains of the acidic amino acids Asp and Glu have pK and Lys, have pK values of 3.9 and 4.1 respectively, whereas those of the basic amino acids Arg and Lys, have pK values of 12.5 and 10.8, respectively. Only the side-chain of His, with a pK value of 6.0, is ionized within the physiological pH range (pH–6.8). It should be borne in mind that when the amino acids are linked together in proteins, only the side-chain groups and the terminal α -amino and α -carboxyl groups are free to ionize.

Reactions

The characteristics organic reactions of amino acids are those of the α -carboxyl group, the α -amino group and the functional groups present in different side chains (R). Some of their reactions are given below.

Reactions of the carboxyl group

1. Esterification: The carboxyl group of an amino acid may be esterified with alcohols. This reaction is often used as a means of protecting the carboxyl group of amino acids in the chemical synthesis of peptides.

$$H_2N\text{–}\underset{\displaystyle R}{\underset{|}{CH}}\text{–}COOH + C_2H_5OH \rightleftharpoons H_2N\text{–}\underset{\displaystyle R}{\underset{|}{CH}}\text{–}\overset{\displaystyle O}{\overset{||}{C}}\text{–}OC_2H_5 + H_2O$$

2. Reduction: The carboxyl group of an amino acid can be reduced with the potent reducing agent lithium borohydride to yield the corresonding primary alcohol. This reaction is useful in the study of amino sequence.

$$H_2N\text{–}\underset{\displaystyle R}{\underset{|}{CH}}\text{–}COOH \xrightarrow{LiBH_4} H_2N\text{–}\underset{\displaystyle R}{\underset{|}{CH}}\text{–}CH_2OH$$

3. Sorenson's formal titration: The carboxyl group of an amino acid cannot be directly titrated against alkali because of the formation of dipolar ion. In the presence of excess of neutralized formaldehyde, the amino acid solution can be directly titrated against alkali as the amino group which interferes with carboxyl group is Converted into a mono–and subsequently into a dimethylol derivatives of an amino acid.

$$HCHO + H_2NCHCOO^- \longrightarrow HOH_2CHNCHCOO^-$$
(with R on the CH in each)

N–Hydroxymethyl derivative

$$HOH_2CHNCHCOO^- + HCHO \longrightarrow HOH_2CNCH{-}COO^-$$
(R on the CH of the reactant; R and CH2OH on the N of the product)

N, N–Bishydroxymethy derivative

Reaction of amino group

1. Reaction with nitrous acid (**Van slyke reaction**): The amino group of an amino acid will react with the strong oxidizing agent nitrous acid (HNO_2) to liberate N_2. This reaction which is stoichiometric is important in the estimation of α - amino groups in amino acids. The amino acids proline and hydroxyproline do not react. The products are the corresponding α - hydroxy acid and N_2 gas, which can be measured manometrically.

$$R{-}\underset{NH2}{CH}{-}COOH + HNO_2 \longrightarrow R{-}\underset{OH}{CH}{-}COOH + N_2 + H_2O + H^+$$

2. Reaction with ninhydrin: The amino group of amino acids on heating with ninhydrin (oxidizing agent) undergoes oxidation to form ammonia, CO_2 and the aldehyde:

$$R{-}\underset{NH_2}{CH}{-}COOH^+ \text{ oxidized ninhydin} \longrightarrow$$

$$R{-}\underset{\overset{||}{O}}{CH} + NH_3 + CO_2 + \text{reduced ninhydrin}$$

A second equivalent of ninhydrin (oxidized) then reacts with the reduced ninhydrin and NH_3 formed in above equation to produce a highly coloured product (Ruheman's purple).

Oxidized ninhydrin + NH_3 + Reduced ninhydrin (hydrindatin) → Purple pigment + $3H_2O$

This reaction is very widely used to estimate amino acids quantitatively in very small amounts. A purple colour is given in the ninhydrin reaction by all the amino acids and peptides having a free α - amino group, whereas, proline and hydroxyproline in which the α - amino group is substitude (imino acids) yield derivative with a characteristic yellow colour.

3. Reaction with 1–fluoro–2,4–dinitrobenzene (FDNB): The compound FDNB (Sanger's reagent) reacis with the free amino group on the NH_2–terminal end of a polypeptide as well as the amino groups of free amino acids and produces the bright yellow coloured dinitrophenyl (DNP) – amino acid derivative. This reaction can be used to identify the N–terminal amino acids of a peptide or protein.

4. Reaction with dansyl chloride (5–dimethylamino–naphthalene–1–sulfonyl chloride): The amino group of both free amino acids and peptide chains will react with dansyl chloride to produce a dansyl amino acid derivative which fluoresces under UV light. This reaction may be also used to find our N–terminal amino acid of a peptide or protein.

$$H_2N-CH(R)-CO_2H + NO_2-C_6H_3(NO_2)-F \longrightarrow NO_2-C_6H_3(NO_2)-NH-CH(R)-CO_2H + HF$$

FDNB DNP-Amino acid

5. Reaction with phenylisothiocyanate (Edman's reaction): Phenylisothiocyanate reacts with α - amino group of an amino acid (or peptide) to form the corresponding phenylthiocarbamyl amino acid. In anhydronus acid this compound cyclizes to form a phenylthiohydration which is stable in acid. This reaction is used both to degrade a polypeptide chain and to identify the N–terminal amino acid in the peptide.

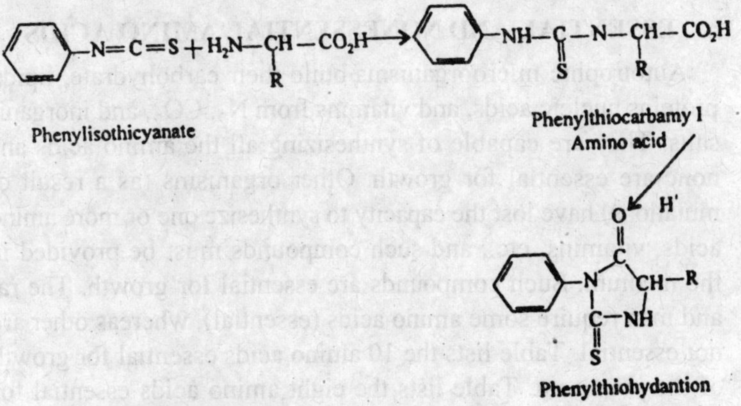

$$C_6H_5-N=C=S + H_2N-\underset{R}{\underset{|}{CH}}-CO_2H \rightarrow C_6H_5-NH-\underset{S}{\underset{\|}{C}}-N-\underset{R}{\underset{|}{CH}}-CO_2H$$

Phenylisothicyanate

Phenylthiocarbamy l Amino acid

H^+

Phenylthiohydantion

Reactions of the R group

Amino acids also show qualitative colour reactions typical of certain functions present in their R groups,e,g. the thiol group of cysteine, the phenolic hydroxyl group of tyrosine and the guanidinium group of arginine.

1. Oxidation of cysteine to cystine:

The thiol group of cysteine is highly susceptible to atmospheric oxygen in the presence of iron salts or by other mild oxidizing agents. The oxidation product is cystine in which the disulfide bond constitutes a covalent bridge between two residues of cystine. Cystine plays an important role in protein structure since its disulfide group serves as a covalent cross link between two polypeptide chains or between two points as a single chain.

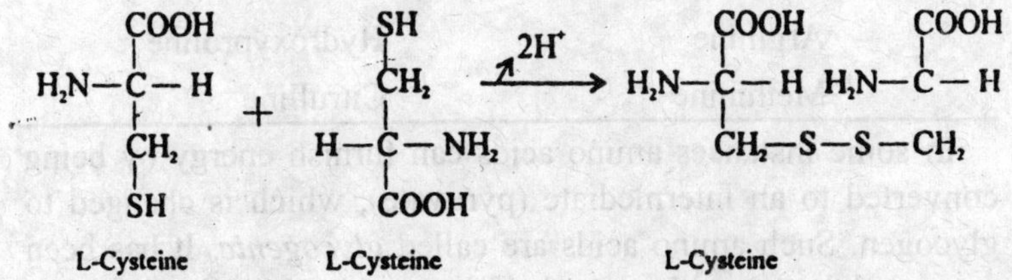

$$H_2N-\overset{COOH}{\overset{|}{C}}-H \quad (CH_2-SH) + H-\overset{CH_2-SH}{\overset{|}{C}}-NH_2 \quad (COOH) \xrightarrow{2H^+} H_2N-\overset{COOH}{\overset{|}{C}}-H \quad H_2N-\overset{COOH}{\overset{|}{C}}-H \quad CH_2-S-S-CH_2$$

L-Cysteine L-Cysteine L-Cysteine

2.Reaction of cysteine with Ag^+: The thiol group of cysteine reacts with heavy metal ions such as Hg^{2+} and Ag^+ to form mercaptides.

ESSENTIAL AND NONESSENTIAL AMINO ACIDS

Autotrophic microorganisms build their carbohydrate, lipids, proteins nucleic acids, and vitamins from N_2, CO_2, and inorganic salts. They are capable of synthesizing all the amino acids and none are essential for growth. Other organisms (as a result of mutations) have lost the capacity to synthesize one or more amino acids, vitamins, etc., and such compounds must be provided in the medium. Such compounds are essential for growth. The rat and man require some amino acids (essential), whereas other are not essential. Table lists the 10 amino acids essential for growth of the young rat. Table lists the eight amino acids essential for the maintenance of nitrogen equilibrium in young adult men:

TABLE. 2.1 CLASSIFICATION OF AMINO ACIDS WITH RESPECT TO THEIR GROWTH EFFECTS IN THE RAT

Essential	*Nonessential*
Lysine	Glycine
Troptophan	Alanine
Histidine	Serine
Phenylalanine	Cystine
Leucine	Tyrosine
Isoleucine	Aspartic acid
Threonine	Glutamic acid
Valine	Proline
Arginine	Hydroxyproline
Methinine	Citrulline

In some instances amino acids can furnish energy by being converted to an intermediate (pyruvate), which is changed to glycogen. Such amino acids are called *glycogenic.* It has been estimated that more than one half the protein N of an animal is a potential source of carbohydrate. On the other hand, phenylalanine, tyrosine, leucine, and isoleucine yield products that belong to the ketone bodies, and are called *ketogenic.* In the ketogenic group, phenylalanine, tyrosine, and isoleucine are both glycogenic and ketogenic.

TABLE. 2.2: MINIMUM AND RECOMMENDED INTAKE FOR NORMAL MAN WHEN DIET FURNISHES SUFFICIENT NITROGEN FOR SYNTHESIS OF NONESSENTIAL AMINO ACIDS

Amino acid *Recommended Daily*	*Rqt. (gm.)*	*Minimum Daily* *Intake(gm.)*
L–Tryptophan	0.25	0.5
L–Phenylalanine	1.10	2.2
L–Lysine	0.80	1.6
L–Threonine	0.50	1.0
L–Valine	0.80	1.6
L–Methionine	1.10	2.2
L–Leucine	1.10	2.2
L–Isoleucine	0.70	1.4

Nitrogen fixation and assimilation

Atmospheric nitrogen serves as the source of plant nutrition and, in the form of ammonia, for animal nutrition. Microrganisms in the soil fix nitrogen to form ammonia. Nitrates are the most useful form of nitrogen for crop plants. Among the organisms which carry out nitrogen fixation are the autotrophs *Rhodospirillum rubrum*, a bacterium *Nostoc muscorum*, a blue–green alga, and *Azotobacter* and *Clostridium pasteuranum*, which are bacterial heterotrophs. Symbiotic nitrogen fixation takes place in nodules on the roots of leguminous plants by *Rhizobium;* neither the roots nor the bacteria can perform this function alone.

In cell–free systems from microorganisms, pyruvate plays an important and dual role: it provides a source of ATP and it can also act as a reducing agent. As pyruvate is oxidized, the iron–

protein *ferredoxin* is reduced, transferring its reducing power to the *nitrogenase* system which catalyzes the conversion of N_2 to ammonia:

$$N_2 + 3H_2 \rightarrow 2NH_3$$

Little is known of the mechanism of N_2 activation by the enzyme, except for the concept of a redox catalyst which couples ATP hydrolysis and electron transfer in the presence of a divalent metal cation, probably Mg^{++}. Bacterial ferredoxin and flavodoxin, a flavin–containing protein, are capable of acting in the nitrogenase system. The enzyme also catalyzes a wide variety of reduction reactions of substrates, including N_2O, N_3^-, alkynes, HCN, nitriles, and isonitriles. However, the enzyme has a highest affinity for N_2.

Nitrosification refers to the oxidation of ammonia to nitrite. An example of a bacterium that carries on this reaction is *Nitromonas*. *Nitrification* is the oxidation of nitrites to nitrates. An example of a bacterium with this function is *Nitrobacter*. The bacteria involved in these oxidations are autotrophic— oxidize ammonia to nitrites ($\Delta F° = -66{,}500$ cal.) and nitrites to nitrates ($\Delta F° = -17{,}500$ cal.) as a source of energy. This energy is utilized for the reduction of CO_2 to carbohydrate and other compounds of carbon.

Higher plants microorganisms make use of nitrates in the soil by first reducing it to ammonia. *Nitrate reductase* catalyzes the reduction of nitrate to nitrite, and *nitrite reductase* further reduces nitrites to ammonia. The latter reaction may involve as intermediate hyponitrite and hydroxylanine. The enzyme *nitrate reductase* is a pyridine–dependent molybdoflavoprotein, and the *nitrite reductase* is also a pyridine–dependent metalloflavoprotein. In both instances ferredoxin plays a central role as an electron source. The ammonia which is formed can now serve as a source of nitrogen for amino acids in plants and animals.

Fixation of ammonia as glutamic acid and glutamine

The enzyme *glutamic acid dehydrogenase* catalyzes the synthesis as well as the oxidation of glutamic acid. The equilibrium of the reaction favours glutamic acid synthesis:

$$\alpha\text{ - Ketoglutaric acid} + H^+ + NH_3 \rightleftarrows \text{L–glutamate} + H_2O$$

Whereas the enzyme from plants and most animal tissues makes use of NAD^+, either NAD^+ or $NADP^+$ is utilized by the liver enzyme. One preparation of the enzyme has a molecular weigth of 250,000 and is make up of 5 subsunits of 50,000 each. These subunits may be held together by Zn^{++}. This reaction may well be the most important quantitatively for fixation of ammonia.

The enzyme *glutamine synthetase* catalyzes the fixation of ammonia by glutamic acid in the form of the amide group of glutamine:

$$\text{L–Glutamic acid} + ATP + NH_3 \rightleftarrows \text{L–glutamine} + ADP + P_i + H_2O$$

This is an important storage form of ammonia, which may be very toxic in many organisms, as well as a specific source of ammonia required in many syntheses, e.g., purines and pyrimidines. The enzyme has a molecular weight of 592,000 and consists of 12 subunits of 48,500 each. Two forms of the enzyme have been isolated: *synthetase I,* the most active species, and *synthetase II,* which is much less active and contains covalently bound AMP. The enzyme *adenyl transferase* is responsible for the conversion of I to II:

$$\text{Synthetase I} + ATP + Mg^{++} \rightleftarrows \text{Synthetase II}$$

This constitutes a system for regulating glutamine synthesis; another enzyme, *glutaminase,* catalyzes the hydrolytic release of ammonia from glutamine. Synthetase II can be deadenylated in a reaction involving as yet unpurified enzymes, α - ketoglutaric acid and a mixture of UTP and ATP. This reaction is inhibited by glutamine.

Amino Acid Nitrogen Interchange; The Nitrogen Pool

Transaminases

The presence of enzymes that catalyze the transfer of the amino group from one amino acid to an α -keto acid to form new amino acid and corresponding. The transfer of sulfate from PAPS to an acceptor –e.g., phenols, chondroitin, dehydroandrosterone, and probably heparin–is catalyzed by specific enzymes.

The anaerobic degradation of cysteine results in the release of H_2S, pyruvic acid and NH_3:

$$H_2O + \underset{\text{Cysteine}}{\begin{array}{c} CH_2SH \\ | \\ CH{\cdot}NH_2 \\ | \\ COOH \end{array}} \longrightarrow \underset{\text{Pyruvic acid}}{\begin{array}{c} CH_3 \\ | \\ C{=}O \\ | \\ COCH \end{array}} + H_2S + NH_3$$

Sulfide is removed from the body in the from of S_2O_3, which is converted by small amounts of CN^- to SCN^-:

$$S_2O_3^{=} + CN^- \rightarrow SO_3^{=} + SCN^-$$

The enzyme is called rhodaneseor $S_2O_3^{=}$–transsulfurase.

Cysteine is a precursor of taurine, which is conjugated with cholic acid (a bile acid) to form taurocholic acid, which at neutral pH, is one of the bile salts. The other is glycocholate. This involves the intermediate formation of cholyl CoA:

$$\text{Cholic acid} + \text{CoA} + \text{ATP} + Mg^{++} \rightarrow \text{cholyl CoA}$$

Cholyl CoA + taurine → [structure: OH, HO, OH, $CO{\cdot}NH{\cdot}CH_2{\cdot}CH_2{\cdot}SO_3^-$]

Taurocholic acid

[bromobenzene, Br] + Cysteine + "Acetic acid" → [$S{\cdot}CH_2{\cdot}CH{\cdot}COOH$, $NH{\cdot}CO{\cdot}CH_3$, Br]

p-Bromphenylmercapturic acid

Cysteine takes part in reactions with compounds foreign to the animal when these compounds are introduced into the body. An example is the formation of a mercapturic acid (acetyl cysteine) derivative of brombenzene.

Formation of Urea

The average human excretes 30 gm of urea in 24 hours. This compound has always been regarded as the end product of nitrogen (protein) metabolism. In some species, urea is replaced by uric acid or ammonia.

Early explanations for the formation of urea in the body based on liver perfusion experiments assumed an overall reaction involving ammonia (obtained by the deamination of amino acids) and carbon dioxide present in blood:

$$2NH_3 + CO_2 \rightarrow CO\,(NH_2)_2 + H_2O$$

TABLE 2.3: END PRODUCTS OF PROTEIN AND PURINE METABOLISM IN VARIOUS VERTEBRATES, CORRELATED WITH THE PRESENCE OF LIVER ARGINASE

	End Product of		
	Protein Metabolism	*Purine Metabolism*	*Liver Arginase*
Mammalia	Urea	Allantoin	+
Birds	Uric acid	Uric Acid	–
Reptilia:			
Snakes, lizards	Uric acid	Uric acid	–
Turtles	Urea	Allantoin	+
Amphibia	Urea	Urea	+
Pisces:			
Elasmobranchii (sharks, dogfish, etc.)	Urea	Urea	+
Telesotei (most bony fish)	Ammonia	Urea	+

Later Krebs and Henseleit, utilizing the tissue slice technique, showed that ammonium salts could be converted to urea. Of the many substances tested for their possible stimulating effect on this reaction, ornithine and citrulline showed striking activity. It was discovered that the action of ornithine is a catalytic one; only small quantities were needed, and there was no stoichiometric relation between the quantity of ornithine added and the amount of urea formed.

In an attempt at an explanation, Krebs assumed the formation of an intermediate compound—by the interaction of ornithine, carbon dioxide, and ammonia—which then broke down to form urea. This intermediate compound he showed to be arginine, which in the presence of the enzyme *arginase* forms urea.

In this connection it is important to note that there is a correlation between the ability of a tissue to form urea and its content of arginase. In animals that excrete urea, arginase occurs chiefly in the liver.

It was postulated that citrulline was doubtless an intermediate in the conversion of ornithine to arginine. Citrulline, originally isolated from the seeds of the watermelon, has since been shown to occur in animals.

The theory of Krebs can be summarized as follows:

$$\underset{\text{Ornithine}}{\begin{array}{l} CH_2NH_2 \\ | \\ CH_2 \\ | \\ CH_2 \\ | \\ CH(NH_2) \\ | \\ COOH \end{array}} \xrightarrow[-H_2O]{CO_2 + NH_3} \underset{\text{Citrulline}}{\begin{array}{l} O{=}C{-}NH_2 \\ \quad | \\ CH_2NH \\ | \\ CH_2 \\ | \\ CH_2 \\ | \\ CH(NH_2) \\ | \\ COOH \end{array}} \xrightarrow[-H_2O]{NH_3} \underset{\text{Arginine}}{\begin{array}{l} HN{=}C{-}NH_2 \\ \quad | \\ CH_2NH \\ | \\ CH_2 \\ | \\ CH_2 \\ | \\ CH(NH_2) \\ | \\ COOH \end{array}} \xrightarrow[\text{(arginase)}]{H_2O} \begin{array}{l} \underset{\text{Urea}}{H_2N{-}\overset{O}{\overset{\|}{C}}{-}NH_2} \\ + \\ CH_2NH_2 \\ | \\ CH_2 \\ | \\ CH_2 \\ | \\ CH(NH_2) \\ | \\ COOH \\ \text{Ornithine} \end{array}$$

(Ornithine formed is returned to the start of the cycle.)

What was needed was an explanation of the mechanism of the reactions, the enzymes involved, and the source of energy need for the production of urea.

The mechanism for the biosynthesis of citrulline from ornithine, carbon dioxide, and ammonia involves two enzymatically catalyzed steps. These are:

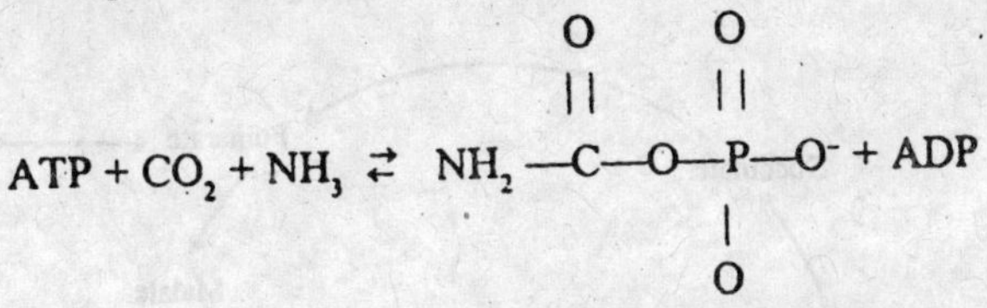

$$ATP + CO_2 + NH_3 \rightleftarrows NH_2-\overset{\overset{O}{\|}}{C}-O-\underset{\underset{O}{|}}{\overset{\overset{O}{\|}}{P}}-O^- + ADP$$

Carbamyl phosphate

Carbamyl phosphate + ornithine ⇄ citrulline + $PO_4^{\equiv}$

In mammalian liver, acetyl glutamic acid or related derivatives of glutamic acid are required as cofactors for the synthesis of carbamyl phosphate.

The mechanism of coversion of citrulline to arginine has been clarified by the studies of Ratner, who separated the enzymes required for this conversion and identified the intermediates. The nitrogen donor is not ammonia, but aspartic acid:

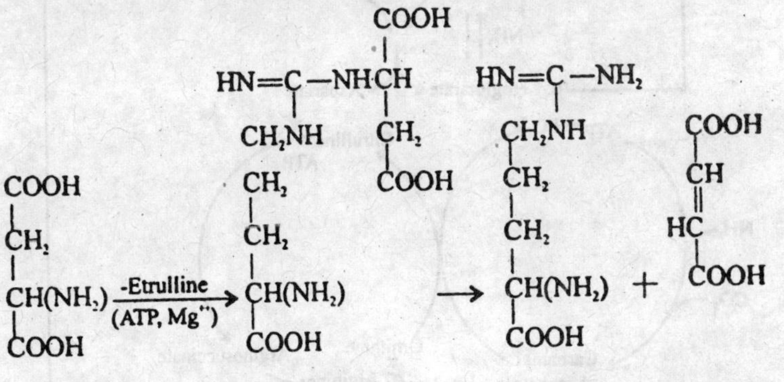

$$\underset{\textbf{Aspartic acid}}{\begin{matrix}COOH\\|\\CH_2\\|\\CH(NH_2)\\|\\COOH\end{matrix}} \xrightarrow[\text{(ATP, Mg}^{++}\text{)}]{\text{-Etrulline}} \underset{\textbf{Argininosuccinic acid}}{\begin{matrix}HN{=}C{-}NH{\cdot}CH(COOH)CH_2COOH\\|\\CH_2NH\\|\\CH_2\\|\\CH_2\\|\\CH(NH_2)\\|\\COOH\end{matrix}} \longrightarrow \underset{\textbf{Arginine}}{\begin{matrix}HN{=}C{-}NH_2\\|\\CH_2NH\\|\\CH_2\\|\\CH_2\\|\\CH(NH_2)\\|\\COOH\end{matrix}} + \underset{\textbf{Fumarate}}{\begin{matrix}COOH\\|\\CH\\\|\\HC\\|\\COOH\end{matrix}}$$

The picture of urea synthesis, as it emerges from this work, indicates that the ammonia utilized for urea synthesis does not arise *per se* from oxidative deamination if amino acids. Ammonia is used in the formation of carbamyl phosphate, and some of it is actually supplied in the form of aspartic acid. Of course, aspartic acid can arise in mammalian tissues from ammonia in the following way:

$$NH_3 + \alpha\text{ - ketoglutarate} \rightarrow \text{glutamic acid}$$

$$\text{Glutamic acid + oxalacetic acid} \rightarrow \alpha - \text{ketoglutarate + aspartic acid}$$

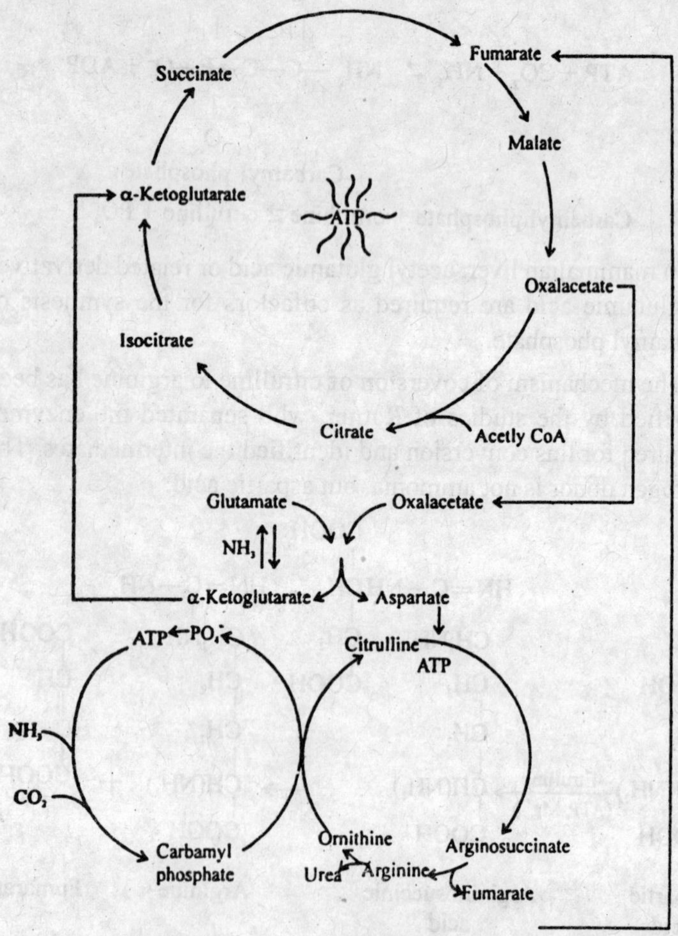

Figure. 2.10: ***Relationship of urea synthesis to the tricarboxylic acid cycle. Two moles of ATP are required for the snthesis of one mole of urea.***

The following illustrates the relationship of urea synthesis to the Krebs cycle and also serves to demonstrate the origin and metabolic fate of the amino acids arginine, glutamic acid and aspartic acid. Other amino acids are related to members of the

Krebs cycle e.g., alanine via pyruvic acid, and phenylalanine (and tyrosine) via its oxidative intermediate fumaric acid. The ketogenic amino acids (e.g., phenylalanine) may yield acetoacetic acid biosynthesis, via acetyl CoA. Aspartic acid is a building block for the pyrimidines and glycine is an important residue in the structures of purines, heme, creatinine, bile salts, etc. Cysteine gives rise to alanine, glutathione, methionine, and choline.

3

Individual Amino Acid Metabolism

Amino acids are of main importance in the metabolism of all organism because they are the precursors of protein. From nutritional aspects, the amino acids may be classified into three groups, namely **dispensable, indispensable** (*essential*) and **semidispensable** amino acids. Dispensable amino acids are those which can be synthesized at the desired rates in the human body out of the substances ordinarily available in the body. Indispensable amino acids are those which can not be synthesized in the body under ordinary circumstances. Some of the indispensable amino acids may be synthesized under certain special circumstances when special ingredients are semidispensable amino acids are those which under ordinary circumstances are synthesized in only very small quantities which do not commensurate with the body needs. Examples of each category are given below:

Amino Acids

Indispensable	**Semidispensable**	**Dispensable**
1. Lysine	1. Arginine	1. Glutamic acid
2. Tryptophan	2. Histidine	2. Aspartic acid
3. Phenylalanine	3. Glycine	3. Alanine
4. Threonine	4. Cystine	4. Proline
5. Valine	5. Tyrosine	5. Hydroxyproline

6. Methionine 6. Serine
7. Leucine
8. Isoleucine

For all practical purposes, it may be assumed that the indispensable amino acids are not synthesized in the body, however, these are absolutely essential for adequate growth, and hence are also known as **essential amino acids.** These must be obtained by the body from diet in the preformed stage. The nutritive value of a particular protein depends upon the number and quantity of essential amino acids it contains. If a given protein is deficient even in a single essential amino acid, its ability to synthesize body proteins is considerably reduced. Such proteins are merely used by the body as source of energy. For positive nitrogen balance, it is absolutely necessary that all the essential amino acids should be obtained by the body in adequate amounts at the same time. It is not necessary that a single protein should contain all the essential acids. These may be obtained from a **mixture** of several proteins, derived from different sources. Since amino acids are not stored as such in the body, administration of the missing essential amino acids afterwords does not serve the purpose.

SYNTHESIS OF ESSENTIAL AMINO ACID

As pointed out already that an amino acid is classed as essential if it must be supplied in the diet for optimal growth of an immature animal or for the maintenance of nitrogen balance in the mature animal. In the context of this definition histidine may be considered borderline as a dietary essential for human beings because it can be synthesized and is also salvaged from protein breakdown to a considerable degree. However, since it is essential for infants, this amino acid will be included in the group of nine amino acids to be described in this chapter.

As is true for the nonessential amino acids, the biochemical reactions of the essential amino acids are integrated with the pathways of carbohydrate and lipid metabolism. The interplay of bidirectional and unidirectional processes in these metabolic

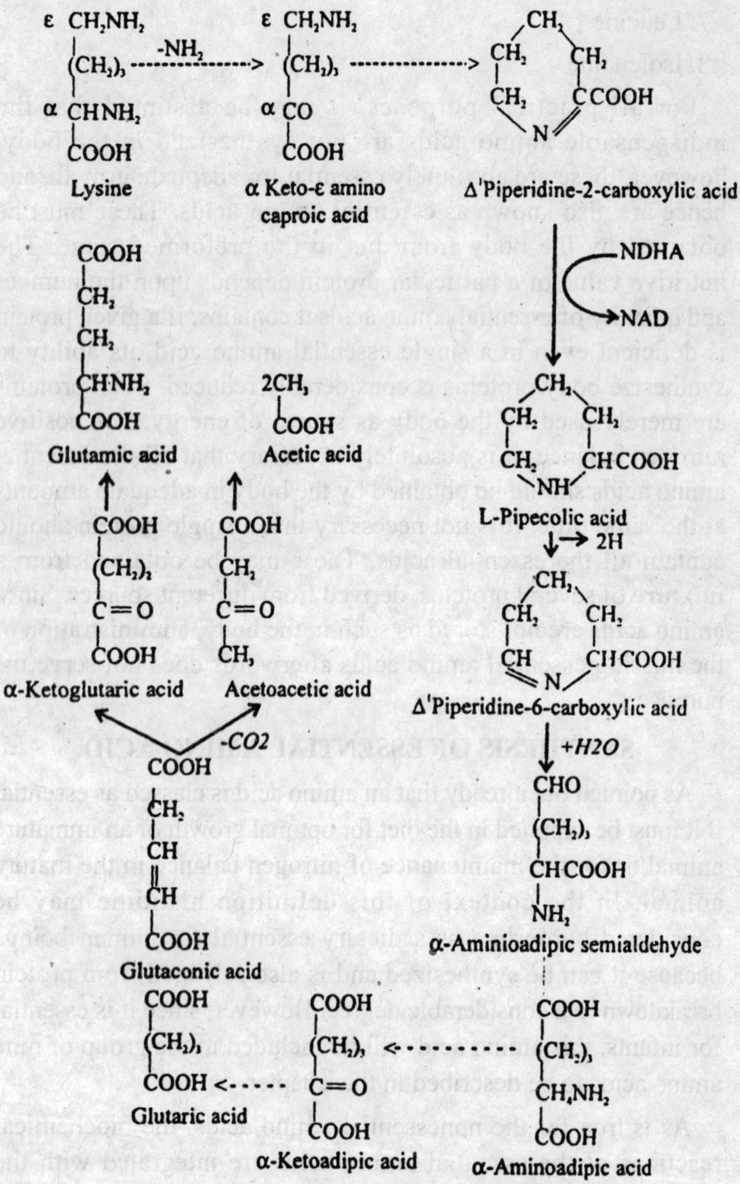

Figure. 3.1 : ***Lysine metabolism in the liver tissue.***

relationship may be viewed as follows:

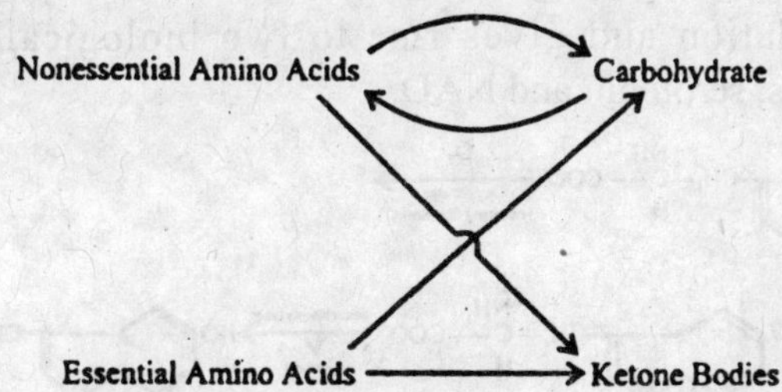

Inborn errors of human metabolism were first recognized in the area of amino acid metabolism. Although catabolic reactions of both the essential and nonessential amino acids are affected by such genetic defects. The inherited errors in metabolism of essential amino acids such as phenylalanine, tryptophan, and the branched chain acids are particularly prominent. We now examine the major catabolic pathways of the essential amino acids and the precursor roles of some of their metabolic products.

LYSINE

Lysine can not be synthesized in man or other animals. It can not participate in the transamination reactions in the mammals. D–Isomer of lysine can not be utilized. L–Amino acid oxidases of animals origin do not attack lysine unless its ϵ amino group is protected by N–methylation, or N–acetylation.

On the basis of isotopic studies using rat and guinea–pig liver, lysine has been shown to be converted into glutaric acid through a series of reactions which may subsequently be converted into acetic acid or glutamic acid.

In mammals, **hydroxylysine** which is derivative of lysine has been shown to be present in collagen. Hydroxylysine is believed to arise as a result of oxidation of lysine after it has been incorporated in the collagen molecule.

D–lysine can not replace *L*–lysine, however, certain substituted lysine dervatives such as ϵ-methyllysine, or ϵ-acetyllysine can replace dietary lysine for growth in the young rats.

TRYTOPHAN

This essential, glucogenic amino acid has two major pathways of degradation and gives rise to two biologically active compounds, serotonin and NAD.

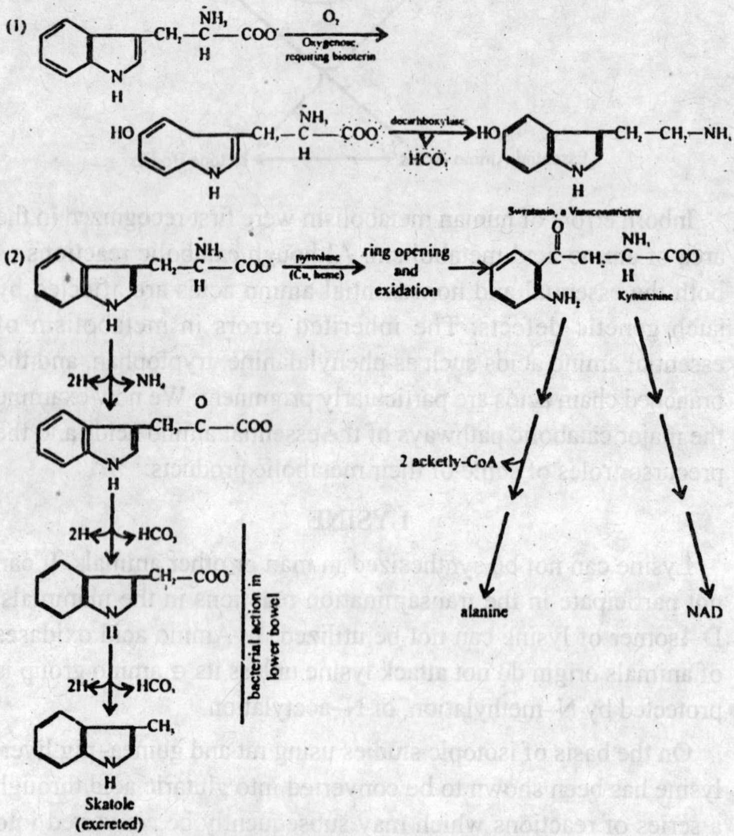

(1) Oxygenase, requiring biopterin; decarboxylase; Serotonin, a Vasoconstrictor

(2) pyrrolase (Cu, heme); ring opening and oxidation; Kynurenine; 2 acketly-CoA; alanine; NAD

bacterial action in lower bowel; Skatole (excreted)

The oxygenase system responsible for the indroduction of an oxygen atom into the tryptophan ring requires the pteridine derivative, tetrahydrobiopterin:

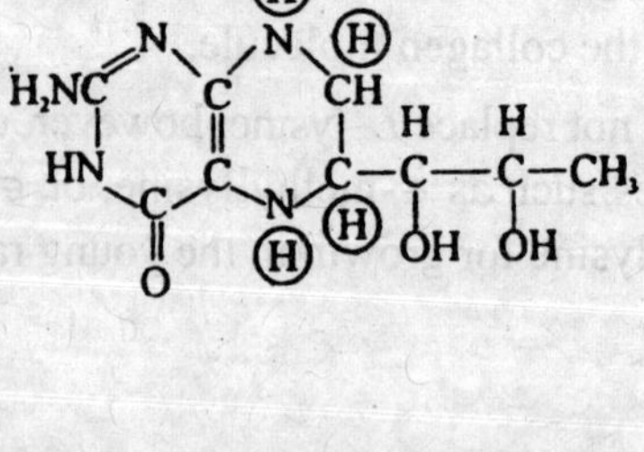

In the oxygenation of the tryptophan the tetrahydrobiopterin is converted to the dihydroform ($BioH_2$):

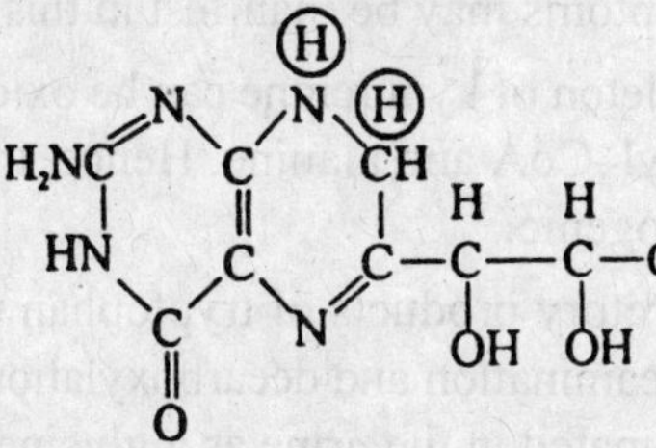

The overall reaction can be summarized schematically as follows:

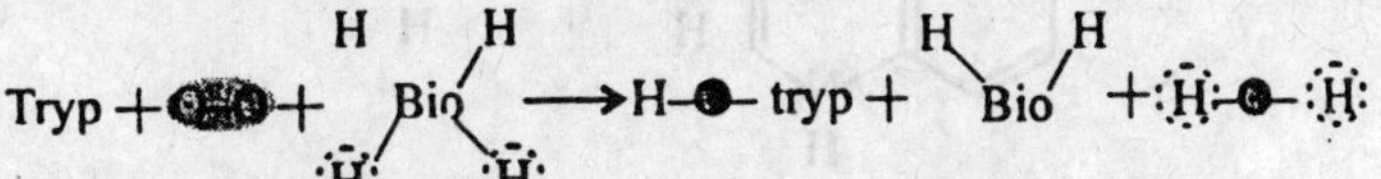

The dihydrobiopterin produced in this reaction can be restored to its tetrahydro form by an NADPH–dependent reduction:

$$BioH_2 + NADPH + H^+ \rightarrow BioH_4 + NADP^-$$

The decarboxylation of the hydroxytryptophan, a reaction prominent in both brain and kidney, yields serotonin. This neurohumoral agent, which is present in brain, platelets, mast cells, and intestinal cells,is a potent vasoconstrictor.

The other major pathway of tryptophan degradation results in its oxidation and ring opening to kynurenine. This pathway, which is prominent in liver, involves as copper– heme enzyme system, termed tryptophan pyrrolase or trytophan 2,3–dioxygenase. Kynurenine can be converted in the liver to quinolinic acid

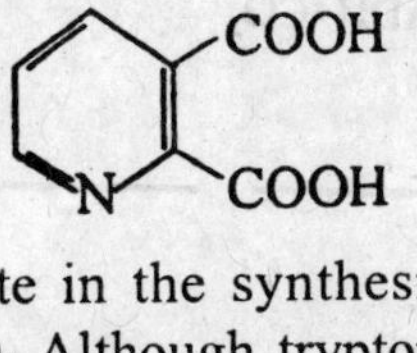

which is an intermediate in the synthesis of nicotinic acid ribonucleotide and NAD. Although tryptophan cannot replace niacin obtained from oxogenous sources, it can have a demonstrable 'sparing' effect on the dietary requirements for the vitamin. HARTNUP'S DISEASE is a hereditary disorder in which there is a deficiency of the pyrolase. In addition to more serious consequences such as mental retardation, it is predictable that pellagra–like symptoms may be manifest in this disease.

The carbon skeleton of kynurenine can be oxidatively degraded to yield both acetyl–CoA and alanine. Hence, tryptophan is both ketogenic and glcogenic.

The major excretory products of tryptophan metabolism arise from successive deamination and decarboxylation reaction. Indole acetic acid is eliminated in the urine as a glycine conjugate.

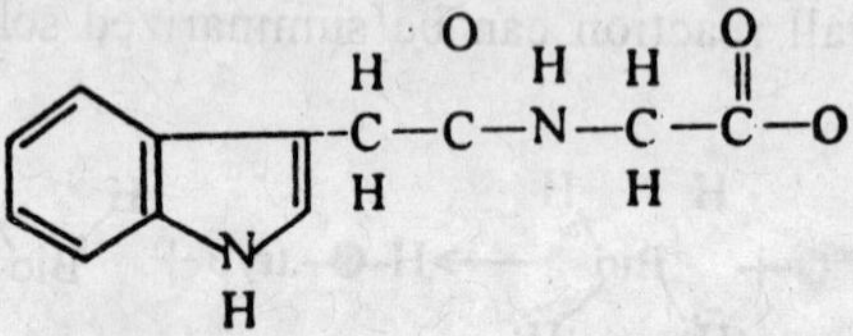

H O H H O
–C–C–N–C–C–O
H H
N
H

More extensive degradation, resulting in skatole, occurs by bacterial action in the lower bowel.

THREONINE

Threonine is a gluconeogenic amino acid. It forms part of most of the proteins. The conversion of threonine of α - oxobutyrate has been

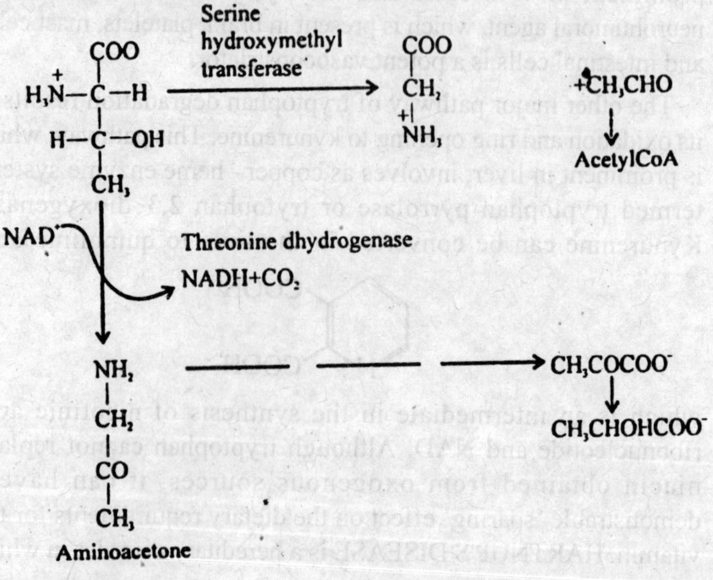

COO
$H_3\overset{+}{N}$—C—H
H—C—OH
CH_3

Serine hydroxymethyl transferase

COO
CH_2
$\overset{+}{N}H_3$

+CH_3CHO
↓
AcetylCoA

NAD^+
Threonine dhydrogenase
NADH+CO_2

NH_2
CH_2
CO
CH_3
Aminoacetone

→ CH_3COCOO^-
↓
$CH_3CHOHCOO^-$

described earlier. The conversion of this compound to propionyl CoA is detailed above. It is believed that alternate catabolic routes for threonine exist. Serine hydroxymethyl transferase is supposed to act on threonine also but in the absence of tetrahydrofolate, to form glycine and acetaldehyde. Aldehyde dehydrogenase can convert acetaldehyde to acetate which can be converted to acetyl CoA. If this sequence of reaction is significant, threonine should be classified as a partly ketogenic amino acid. Threonine dehydrogenase and NAD^+ dependent enzyme, converts threonine to aminoacetone. This is converted to pyruvate and lactate.

Methionine : Aspartic acid gets converted to β - aspartyl semialdehyde by involving its phosphate.

$$HOOCCH_2CH(NH_2)COOH + ATP \xrightleftharpoons{\beta\text{-asparto kinase}} H_2O_3POOCCH_2(NH_2)COOH + ADP$$

β - Aspartyl phosphate

$$H_2O_3POOCCH_2CH(NH_2)COOH \xrightleftharpoons{NADPH \rightarrow NADP^+} OHCCH_2CH(NH_2)\ COOH + Pi$$

β - Aspartyl semialdehyde

It has been reported that the common intermediate is β - aspartyl semialdchyde in the synthesis of methionine, threonine and isoleucine. Homoserine is the branching point for carryng out the synthesis of these amino acids. Homeserine gets' synthesized by the reduction of aspartyl phosphate in the presence of the enzyme called homoserine dehydrogenase. The dehydrogenase may use either NADH or NADPH as coenzyme.

$$OHCCH_2CH(NH_2)COOH + NADPH + H^+ \rightleftharpoons HOH_2CCH_2CH(NH_2)COOH + NADP^+$$

Homoserine

In higher plants it has been reported that homoserine undergoes condensation with a cysteine molecule to form homocysteine, which yields methionine after methylation. However *E. coli* and *Neurospora crassa,* involve the complex synthesis of homocysteine. Alcoholic group of homoserine undergoes acylation first either with succinyl CoA or acetyl CoA to form either O–acetyl–homoserine.

$$HOH_2CCH_2CH(NH_2)COOH + CH_2COOH \longrightarrow$$

$$\begin{array}{c} | \\ CH_2COOH \end{array}$$

$$HOCC(CH_2)_2CHCOOH$$

$$\begin{array}{c} | \\ NH_2 \end{array}$$

O–Succinyl–homoserine

These products on reacting with cysteine from cystathionine, which then gets hydrolyzed to form homocysteine, pyruvate and ammonia.

$$HOOCCH(NH_2)CH_2SH$$

$$HOOC(CH_2)_2COO(CH_2)_2CHCOOH \longrightarrow$$

$$\begin{array}{c} | \\ NH_2 \end{array}$$

$$HOOCCHCH_2S(CH_2)_2CH(NH_2)COOH \quad H_2O$$

$$\begin{array}{c} | \\ NH_2 \end{array}$$

$$\longrightarrow HSCH_2CH_2CHCOOH + CH_3COCOOH + NH_3$$

$$\begin{array}{c} | \\ NH_2 \end{array}$$

Homocysteine

As in higher plants, homocysteine gets methylated at —SH end and the product of the reaction has been found to be methionine. The methyl donor has been found to be another amino acid. serine.

$$HSCH_2CH_2CH(NH_2)COOH \xrightarrow{\text{Serine}} CH_3SCH_2CH_2CH(NH_2)COOH$$

Methionine

LEUCINE

Leucine can undergo transamination reaction forming α - **ketoiso-caproic acid** which is oxidatively decarboxylated in the presence of NAD and CoA.SH to form **isovaleryl coenzyme A** which is an important precursor for synthesis of branched–chain fatty acids. Isovaleryl coenzyme A on further oxidation is coverted into β - **methylcrotonyl coenzyme** A. Carboxylation of this product gives β **-methylglutaconyl coenzyme A.** A molecule of water is then added across the double bond forming β - **hydroxy-β -methylglutaryl coenzyme A** (*HMG–CoA*). This is an important compound which is involved in the biosynthesis of cholesterol and ketone bodies. HMG–CoA is finally degraded into acetoacetic acid and acetyl coenzyme A.

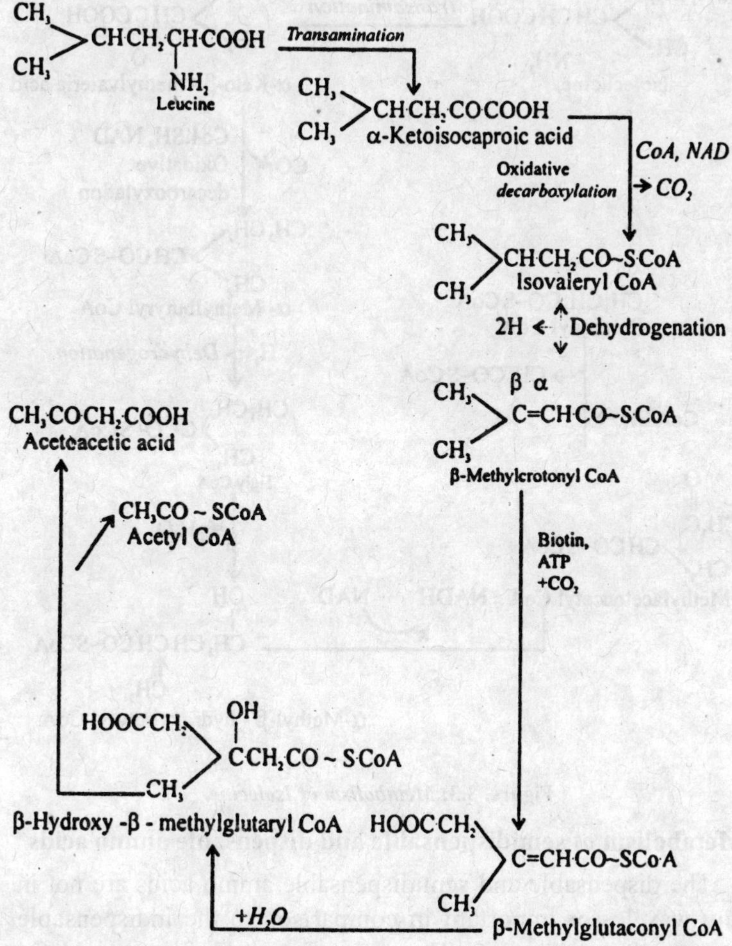

Figure. 3.2: ***Metabolism of Leucine***

Leucine is a **ketogenic amino** acid.

A block in the metabolism of leucine is an inborn error of metabolism and in infants causes a disease known as *maple syrup urine* disease. In this disease, oxidative decarboxylation of α - keto–150–carpic acid is blocked, and hence, it is excreted in the urine in excessive amounts.

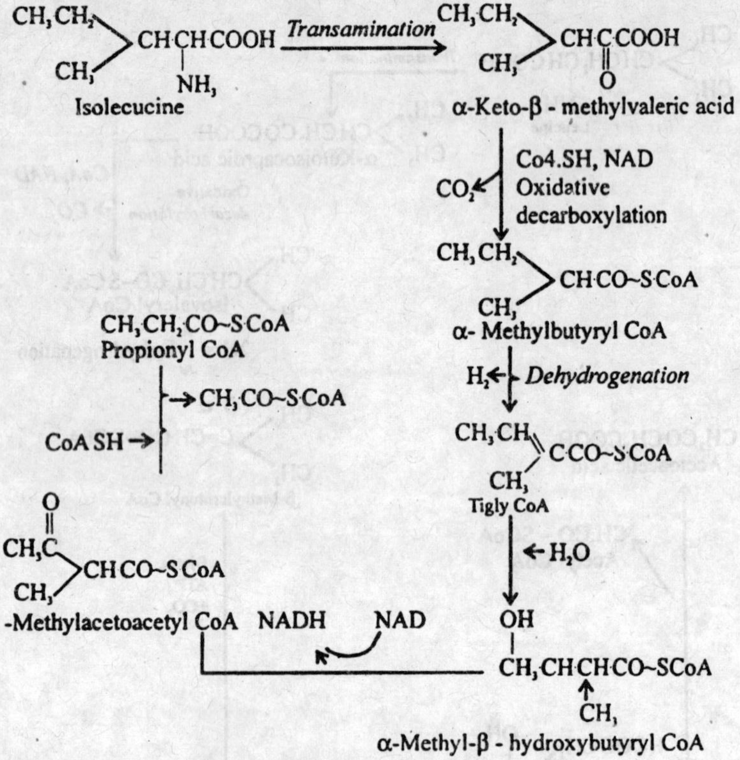

$CH_3CH_2(CH_3)CHCH(NH_2)COOH$ Isolecucine —*Transamination*→ $CH_3CH_2(CH_3)CHC(=O)COOH$ α-Keto-β - methylvaleric acid

↓ Co4.SH, NAD; CO_2 ← Oxidative decarboxylation

$CH_3CH_2(CH_3)CHCO{\sim}SCoA$ α- Methylbutyryl CoA

↓ H_2 ← Dehydrogenation

$CH_3CH{=}C(CH_3)CO{\sim}SCoA$ Tigly CoA

↓ ← H_2O

$CH_3CH(OH)CH(CH_3)CO{\sim}SCoA$ α-Methyl-β - hydroxybutyryl CoA

— NAD → NADH →

$CH_3C(=O)(CH_3)CHCO{\sim}SCoA$ -Methylacetoacetyl CoA

CoA SH → $CH_3CO{\sim}SCoA$ + $CH_3CH_2CO{\sim}SCoA$ Propionyl CoA

Figure. 3.3: *Metabolism of Isolecine.*

Metabolism of semidispensable and dispensable amino acids

The dispensable and semidispensable amino acids are not in any way lesser important in comparison to the indispensable (*essential*) amino acids. These are also similarly required for protein biosynthesis and formation of other biogenic compounds. The only difference being, body has adopted itself for the biosynthesis of dispensable, or semidispensable amino acids but has as yet failed to adopt itself for synthesis of indispensable amino acids. Such amino acids are also designated by some authors as **synthesizable** amino acids.

SEMIDISPENSABLE AMINO ACIDS

Arginine

Arginine is essential in the diet of *only young ones* but is not essential for adults. It can be synthesized in the human body. However in rats, it is not synthesized at a rate commensurate with the demand. Most important function of this amino acid in the body in addition of protein synthesis is it helps in the formation of urea from ammonia and is converted into ornithine to fix up more ammonia.

Arginine may be regenerated from ornithine through a series of reactions. Arginine is gluconeogenic in nature. It is converted into glucose in *diabetes mellitus.* Besides this, arginine can donate an amidine group to glycine and synthesize creatine.

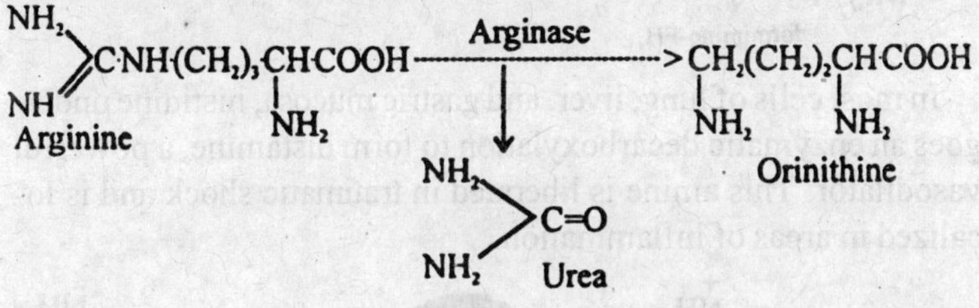

$$\underset{\text{Arginine}}{(NH_2)(NH{=})C{\cdot}NH(CH_2)_3CH(NH_2)COOH} \xrightarrow{\text{Arginase}} \underset{\text{Orinithine}}{CH_2(NH_2)(CH_2)_2CH(NH_2)COOH} + \underset{\text{Urea}}{(NH_2)_2C{=}O}$$

Histidine

Like arginine, histidine can be considered as a 'semis-essential' amino acid. It is a dietary requirement for the young animal but can be synthesized at a rate adequate to maintain nitrogen balance in the adult. In *E.coli* and *Salmonell,* histidine is synthesized from phosphoribosyl–1–pyrophosphate (PRPP), ATP, and glutamine. Its glycogenic character is evident from the given reactions.

A genetic disease, HISTIDINEMIA, manifests itself by a low activity of histidase. In normal metabolism the degradative pathyway initiated by histidae resultse in the formation of a glutamate derivative, formiminoglutamate (FIGLU). The formiminomoiety can be transferred enzymatically to tetrahydrofolate to form N^5–formiminot errahydrofolate. an important ONE–CARBON COMPOUND. In a folate deficiency, FIGLU will be excreted because there is inadequate tetrahydrofolate to "trap" the formimino group.

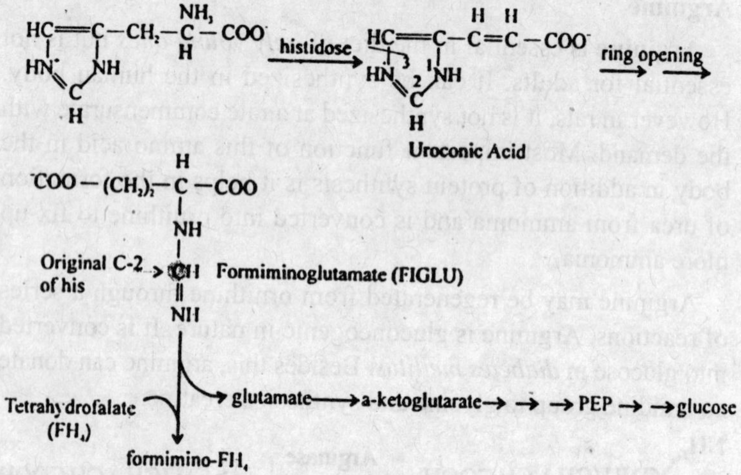

HC=C–CH₂–C(NH₃)(H)–COO⁻ (HN⁺, NH, C, H ring) —histidose→ HC=C–C(H)=C(H)–COO⁻ (HN, NH, C, H ring; 1, 2, 3) —ring opening→

Uroconic Acid

⁻COO–(CH₂)₂–C(H)–COO, NH, Original C-2 of his → CH, NH — Formiminoglutamate (FIGLU)

Tetrahydrofalate (FH_4) → glutamate ⟶ a-ketoglutarate → → → PEP → → glucose

formimino-FH_4

In most cells of lung, liver, and gastric mucosa, histidine undergoes an enzymatic decarboxylation to form histamine, a powerful vasodilator. This amine is liberated in traumatic shock and is localized in areas of inflammation.

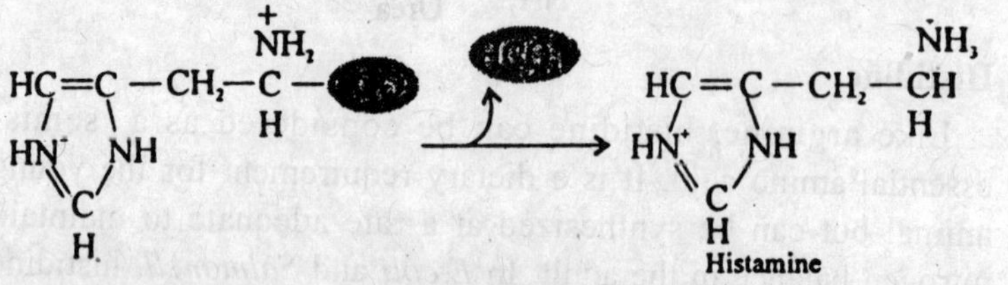

HC=C–CH₂–C(NH₃⁺)(H)–COO⁻ (HN⁺, NH, C, H ring) ⟶ HC=C–CH₂–CH(NH₃)(H) (HN⁺, NH, C, H ring)

Histamine

TYROSINE

Since phenylalanine may be converted into **tyrosine** irreversibly, tyrosine is not an indispensable amino acid. One of the major pathways of phenylalanine metabolism is through its prior conversion to tyrosine, hence, the fate of tyrosine is essentially the same as described under phenylalanine metabolism.

D–Tyrosine can not be metabolized by human beings. **Inability** to utilize tyrosine is observed in the inherited diseases such as *tyrosinosis, alkaptonuria* and *phenylketonuria.*

In *tyrosinosis* the keto–acid of tyrosine., *p*–hydroxyphenylpyruvic acid is excreted in considerable amounts in the urine. However, such patients can utilize homogentisic acid. *Tyrosinosis* is rather a very rare disease. For metabolic details of tyrosine, see phenylalanine metabolism.

Phenylalanine and tyrosine

Humans and most all other animal organisms cannot synthesize the benzene ring. Hence, phenylalanine is a dietary essential. Tyrosine is synthesiszed from phenylalanine. Other than to be utilized for protein synthesis in normal metabolism the only known fate of phenylalanine is its conversion to tyrosine. Much of the dietary requirement for phenylalanine is due, therefore, to the need for tyrosine.

Synthesis of tyrosine

The synthesis of tyrosine from phenylalanine again illustrates a type of synthetic reaction in which molecular oxygen is a direct reactant. The hydroxylation of phenylalanine, like that of tryptophan, requires tetrahydrobiopterin:

$$C_6H_5-CH_2-\underset{H}{\overset{NH_2}{C}}-COO + \text{tetrahydrobiopterin} + O_2 \xrightarrow[\text{hydroxylase}]{\text{phenylalanine}}$$

$$\text{NADPH}+H^+ \rightarrow \text{NADP}^+$$

$$HO-C_6H_4-CH_2-\underset{H}{\overset{NH_2}{C}}-COO + \text{dihydrobiopterin} + HO-H$$

This reaction is the beginning of the metabolism of phenylalanine in normal individuals. In about 1 of 10,000 to 20, 000 newborns, there is a genetic defect because of which phenylalanine hydroxylase is missing or deficient. As a consequene, phenylalanine cannot be converted to tyrosine at a normal rate in these individuals and a second pathway of phenylalanine metabolism, which is normally insignificant, comes into play:

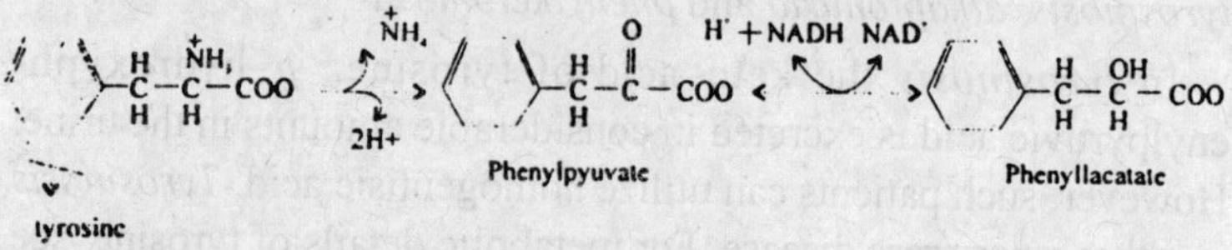

$$\text{Phenylalanine} \xrightarrow[\text{2H}^+]{\text{NH}_4^+} \text{Phenylpyuvate} \xrightarrow[]{\text{H}^+ + \text{NADH} \rightarrow \text{NAD}^+} \text{Phenyllacatate}$$

tyrosine

There is now an increased concentration of phenylalaninen in all body fluids and phenylpyruvate and phenylacetate are excreted in high levels. This disease is called **phenylketonuria.** Early diagnosis of this tragic condition is absolutely essential. A common therapy is a diet low in phenylalanine.

ALKAPTONURIA

The first step in the catabolism of tyrosine is the loss of its α – amino group to form the keto acid.

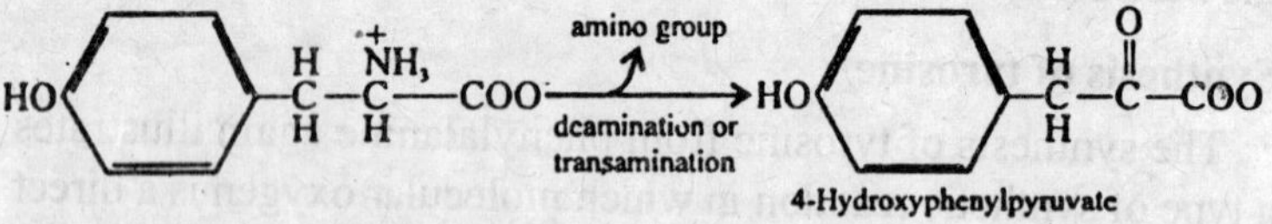

$$\text{HO-C}_6\text{H}_4\text{-CH}_2\text{-CH}(\text{NH}_3^+)\text{-COO}^- \xrightarrow[\text{deamination or transamination}]{\text{amino group}} \text{HO-C}_6\text{H}_4\text{-CH}_2\text{-CO-COO}^-$$

4-Hydroxyphenylpyruvate

The next reaction is a complicated oxidation rearrangement in which another hydroxyl group is added to the ring. The carboxyl group is lost as CO_2, and the positions of the phenyl hydroxyls are changed relative to the side chain. The product of this reaction is homogentisic acid:

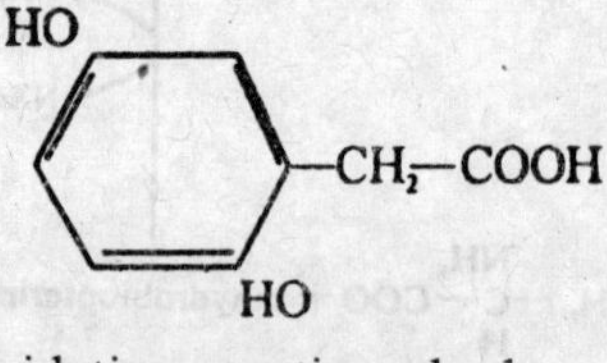

$$\text{(HO)}_2\text{C}_6\text{H}_3\text{-CH}_2\text{-COOH}$$

In a subsequent oxidation reaction, the homogentisic ring is opened, giving an eight –carbon compound:

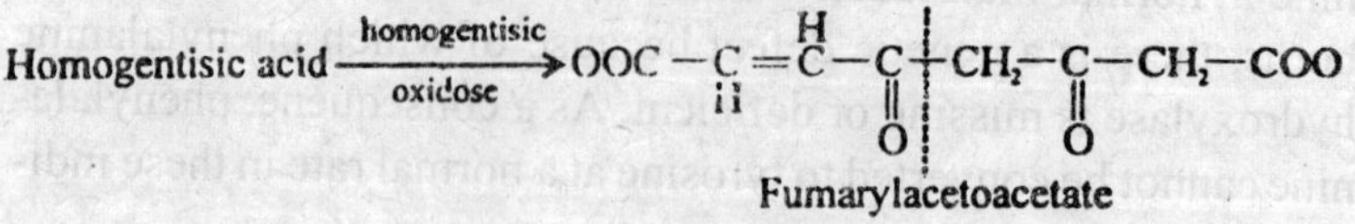

$$\text{Homogentisic acid} \xrightarrow[\text{oxidose}]{\text{homogentisic}} {}^-\text{OOC-CH=CH-CO-CH}_2\text{-CO-CH}_2\text{-COO}^-$$

Fumarylacetoacetate

This compound can be hydrolyzed to fumkarate and acetoacetate:

$$OOC-\underset{H}{C}=\overset{H}{C}-COO+CH_3-\underset{\underset{O}{\|}}{C}-CH_2-COO$$

Phenylalanine and tyrosine are, therefore, both glucogenic and ketogenic.

ALKAPTONURIA is an inherited disorder caused by the absence of homogentisic oxidase, a condition in which homogentisate accumulates and is excreted in the urine. Upon standing, the urine of these individuals turns dark because the homogentisate is further oxidized to a melanin-like compound. Alkaptonuria is generally a benign condition.

Tyrosine as a Precursor of Melanin, Catecholamines, and Thyroxine

Tyrosine can be hydroxylated to yield a derivative that is an important intermediate in mammalian metabolism. In common with other hydroxylation reaction, molecular oxygen and tetrahydrobiopterin ($BioH_4$) are required:

$$HO-C_6H_4-CH_2-\underset{H}{\overset{\overset{+}{N}H_3}{C}}-COO^- \xrightarrow[\text{tyrosine hydroxylase, or tyrosinase}]{BioH_4, O_2, Cu} (HO)_2C_6H_3-CH_2-\underset{H}{\overset{\overset{+}{N}H_3}{C}}-COO^-$$

L-3,4-Dihydroxyphenylalanine (DOPA)

MELANIN is the brownish–black pigment of skin that is producd in the melanocytes in the basal layer of the epidermis. Melanin is a group of high molecular weight polyhydroxyphenyl polymers produced by the action of tyrosinase on DOPA. The oxidation involves two quinone intermediates:

$$DOPA \xrightarrow[\text{tyrosinase}]{(O)} \text{DOPA Quinone} \rightarrow \xrightarrow{(O)} \rightarrow \text{Indole 5,6-quinone} \xrightarrow{(O)} \text{Melanin}$$

DOPA Quinone ($O=C_6H_3(=O)-CH_2-HC(\overset{+}{N}H_3)-COO^-$); Indole 5,6-quinone

A deficiency in the tyrosinase system of the melanocytes results in the condition called *albinism*.

Another important precursor function of **DOPA** is its role in the synthesis of the catecholamines in the adrenal medulla and sympathetic never terminals.

DOPA —DOPA decarboxylase→ HCO_3 + Dopamine

Dopamine —O_2, Dopamine β-hydroxylase (cu,ascorbate)→ Norepinephrine

Norepinephrine —S-adenosylmethionine → S-adenosylhomocysteine→

D(−) Epinephrine, or Adrenalin —S-adenosylmethionine → S-adenosylhomocysteine→ (inactive)

Dopamine β - hydroxylase is a Cu–containing mixed function oxidase. It does not require the tetrahydrobiopterin as in the case of the prior hydroxylation reactions involving the phenyl ring.

Two hormones of the thyroid gland, thyroxine and triiodothyronine, are iodinated derivatives of tyrosine. The overall pathways for the synthesis of these related compounds can be represented as follows:

$HO-C_6H_4-CH_2-CH(\overset{+}{N}H_3)-COO^- + I^- \longrightarrow$ 3-Monoiodotyrosine

+ 3,5-Diiodotyrosine

Thyroxine, or T_4, is formed from two molecules of 3,5–diidotyrosine:

HO–(3,5-diiodophenyl)–O–(3,5-diiodophenyl)–CH_2–C(H)($\overset{+}{N}H_3$)–COO^-

Thyroxine

Triiodothyronine is the product of the reaction of 3–monoiodotyrosine and 3.5–diiodothyronine:

HO–(3-iodophenyl)–O–(3,5-diiodophenyl)–CH_2–C(H)($\overset{+}{N}H_3$)–COO^-

Triiodothyronine, or T_3

T_3 and T_4 are linked to a specific to globulin of the thyroid gland colloid and the combination is called thyroglobulin. Both thyroxine and triiodothyronine must be released from the globulin to exert their hormonal action.

SYNTHESIS OF NON ESSENTIAL AMINO ACIDS

Whether they are dietary essentials or not, all amino acids are required for protein synthesis, all serve as precursors of many compounds that are essential for life, and all can serve as sources of metabolic fuel. Now examine the major biochemical reaction that premit the pathways of amino acid metabolism to be integrated with the pathways of carbohydrate and lipid metabolism. Toward this end, it is helpful to categorize amino acids as glucogenic or ketogenic. If there can be a net synthesis of phosphoenolpyruvate from its carbon skeleton, an amino acid is glucogenic. Hence, all of the amno acids whose carbon chains can be converted into intermediates of the Krebs cycle or to pyruvate, will be glucogenic. Most amino acids are glucogenic. Some amino acids are degraded in such a manner that a part of their carbon skeleton is convertible to phosphoenol pyruvate and the remainder into acetoacetate and other ketone bodies. Therefore, these amino acids are both glucogenic and ketogenic. Some amino acids are completely ketogenic because their intermediate oxidation products can be accounted for entirely as acetyl–CoA and its ketone body derivatives.

Let us now examine the amino acids that are not dietary essentials for humans, accounting for their biosynthesis, their glucogenic or ketogenic character, and their roles as intermediates in the biosynthesis of metabolically essential compounds.

Serine

Serine is a non-essential amino acid because it can be synthesized from glycine and also from 3–**phosphoglyceric** acid, an intermediate of glycolysis

$$\begin{matrix} COO^- \\ HCOH \\ HC{-}O\text{Ⓟ} \\ H \end{matrix} \xrightarrow{2H} \begin{matrix} COO^- \\ C{=}O \\ HC{-}O\text{Ⓟ} \\ H \end{matrix} \xrightarrow[\text{glutamate}]{\text{transamination}} \begin{matrix} COO^- \\ HC{-}\overset{+}{N}H_3 \\ HC{-}O{-}\text{Ⓟ} \\ H \end{matrix} \xrightarrow{P_i} \begin{matrix} COO^- \\ HC{-}\overset{+}{N}H_3 \\ HC{-}OH \\ H \end{matrix}$$

Role in Cysteine Synthesis. The carbon chain of serine can become the carbon chain of the amino acid cysteine:

S-Adenosylmethionine → (CH_3 removed; adenosine released) → $SH{-}(CH_2)_2{-}HC{-}\overset{+}{N}H_3{-}COO^-$ Homocysteine

Homocysteine + serine ($H{-}HO{-}CH{-}HC{-}\overset{+}{N}H_3{-}COO^-$) → $H_2C{-}S{-}CH(H){-}HC{-}\overset{+}{N}H_3{-}COO$ / $H_2C{-}HC{-}\overset{+}{N}H_3{-}COO$ Cystathionine

Cystathionine + HOH → α-Aminobutyric acid + $HS{-}CH(H){-}HC{-}\overset{+}{N}H_3{-}COO$ Cysteine

In humans the formation and splitting of cystathionine are irreversible reactions. Methionine is an essential amino acid. It cannot be made from cysteine.

Function is Complex Lipid Synthesis. One of the major precursor functions of serine is its role in the biosynthesis of the complex lipids. Serine itselfs can be a component of the lipids such as the following:

$$\begin{array}{l} \text{Fatty acyl}-O-\overset{H}{C}H \\ \text{Fatty acyl}-O-CH \quad\quad \overset{O}{\|} \quad\quad\quad\quad\quad \overset{H}{|}\;\; + \\ \quad\quad\quad\quad\quad HC-O-P-O-CH_2-C-NH_3 \\ \quad\quad\quad\quad\quad\; H \quad\quad\;\; | \quad\quad\quad\quad\quad\;\; | \\ \quad\quad\quad\quad\quad\quad\quad\quad O^- \quad\quad\quad\quad COO^- \end{array}$$

Phosphatidlserine

In addition, serine can lose its COOH group to form aminoethanol:

$$HOCH_2-\overset{\overset{+}{N}H_3}{\underset{H}{C}}-COO^- \xrightarrow{HOH} HO-CH_2-\overset{\overset{+}{N}H_3}{\underset{H}{C}}H + HCO_3^-$$

By successive addition of methyl groups to the nitrogen by S–adenosylmethionine, the ethanolamine is converted to choline:

$$CH_3-\overset{CH_3}{\underset{CH_3}{N^+}}-CH_2-CH_2OH$$

All of these reactions can occur with phosphatidyl derivatives.

$$\textbf{Phosphatidylserine} \xrightarrow[CO_2]{} \textbf{phosphatidylethanolamine} \xrightarrow{\cdot CH_3} \xrightarrow{\cdot CH_3} \xrightarrow{\cdot CH_3} \textbf{phosphatidylch}$$

Cysteine

It will be recalled that this amino acid can be synthesized from a reaction between serine and homocysteine.

Cysteine is glucogenic because it can be converted to phosphoenolpyruvate:

Cys + α-KG → HS–CH_2–C(=O)–COO^- + Glu

β-Mercaptopyruvic acid → pyruvate → phosphoenolpyruvate (Sulfide released)

Oxidation of the —SH Group. The oxidizability of the cysteine sulfhydryl group gives the amino acid considerable metabolic versatility and also makes it an important determinant of protein structure:

$$2\text{Cys—SH} \rightarrow \text{Cys—S—S—Cys} + 2\text{H}$$

Thus, both inter–and intrapeptide chain—S—S—linkages can be formed:

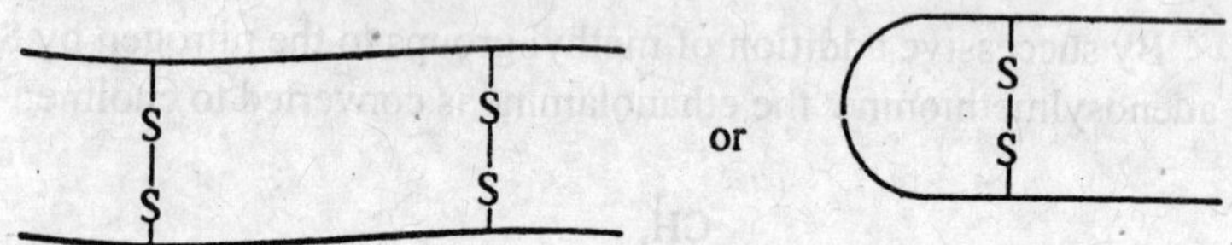

The S of cyseins can also be oxidized beyond the S—S level:

HS–CH_2–HC(–$\overset{+}{N}H_3$)–COO^- → → SO_3H–CH_2–HC(–$\overset{+}{N}H_3$)–COO^- (Cysteic Acid or Cysteine Sulfinic Acid) → (HCO_3^- released) → SO_3H–CH_2–HC(H)–$\overset{+}{N}H_3$ (Taurine)

Cysteic acid → SO_4^{2-} → active sulfate or excreted

Formation of Bile Salts. Bile salts are formed as follows:

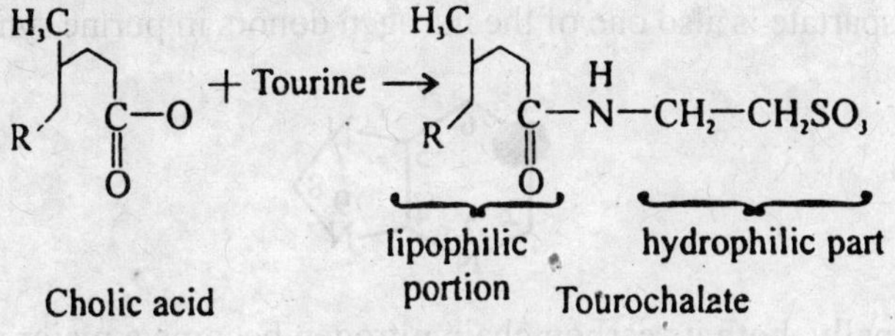

$$\underset{\text{Cholic acid}}{R\text{-}(H_3C)\text{-}C(=O)\text{-}O} + \text{Tourine} \rightarrow \underset{\text{lipophilic portion}}{R\text{-}(H_3C)\text{-}C(=O)}\text{-}\underset{\text{hydrophilic part}}{N(H)\text{-}CH_2\text{-}CH_2SO_3}$$

Tourochalate

Taurocholate is an important bile salt that acts as a detergent in the emulsification process associated with fat digestion and absorption.

Asparate

That this glucogenic amino acid is a dietary nonessential is evident from the following synthetic sequence:

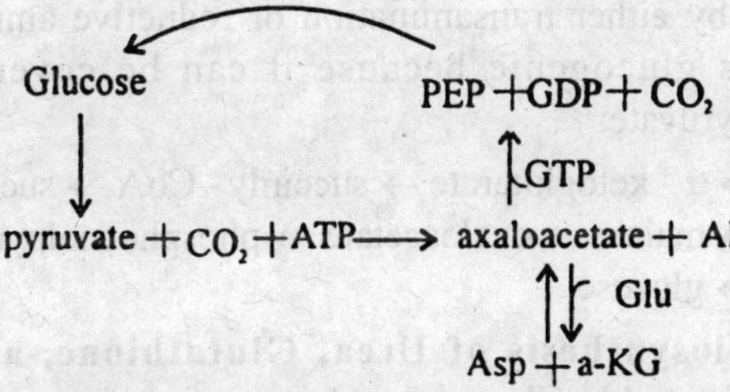

Glucose → pyruvate $+ CO_2 +$ ATP → axaloacetate + ADP $+ P_i$

axaloacetate $\xrightarrow{GTP}$ PEP + GDP + CO_2 → Glucose

axaloacetate $\underset{\text{Glu}}{\rightleftharpoons}$ Asp + a-KG

It should be remembered that the oxaloacetate also serves as the "trigger" for the tricarboxylic acid cycle.

Participation in Synthesis of Urea, Purines, and Pyrimidines. Aspartate participates in three vital synthetic reactions. It contributes one of the nitrogens in the synthesis of urea, serving as a substrate for the synthesis of arginosuccinate:

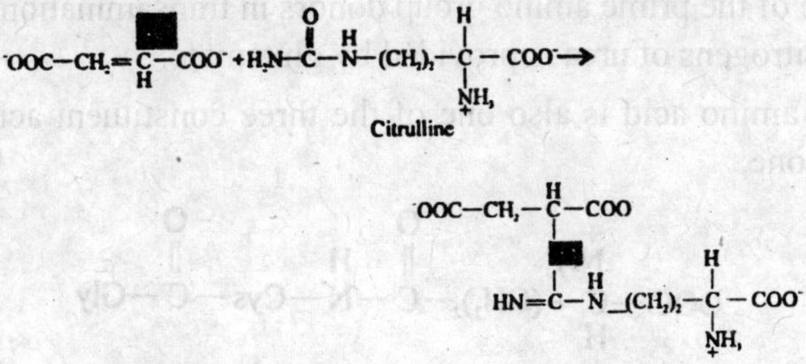

$$^{-}OOC\text{-}CH_2\text{=}\underset{H}{C}\text{-}COO^{-} + \underset{\text{Citrulline}}{H_2N\text{-}\overset{O}{\overset{\|}{C}}\text{-}\overset{H}{N}\text{-}(CH_2)_3\text{-}\overset{H}{\underset{\overset{+}{N}H_3}{C}}\text{-}COO^{-}} \rightarrow$$

$$\begin{array}{c} ^{-}OOC\text{-}CH_2\text{-}\overset{H}{C}\text{-}COO \\ | \\ HN\text{=}C\text{-}\overset{H}{N}\text{-}(CH_2)_3\text{-}\overset{H}{\underset{\overset{+}{N}H_3}{C}}\text{-}COO^{-} \end{array}$$

Aspartate is also one of the nitrogen donors in purine synthesis.

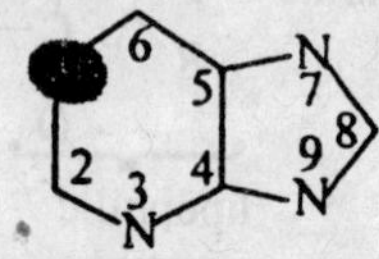

Finally, both its carbon chain nitrogen become a major part of the pyrimidine ring system.

Glutamate

As pointed out previously, this amino acid is formed from α - ketoglutarate by either transamination or reductive amination. Glutamate is glucogenic because it can be coverted to phosphoenolpyruvate:

Glutamate → α - ketoglutarate → succinly –CoA → succinate → fumarate → malate → oxaloacetate → phosphoenolpyruvate (PEP) → → → glucose

Role in Biosynthesis of Urea, Glutathione, and γ - Aminobutyric Acid. Glutamate plays a significant role in biosynthetic reactions and as a precursor of other biological compounds. As pointed out earlier,

$$\text{Glu—}\alpha\text{-NH}_2 \longrightarrow \alpha\text{-KG}$$

$$\overset{+}{NH_4} \xrightarrow{HCO_3 \quad ATP \quad HOH} \text{carbamoyl phospltate}$$

it is one of the prime amino group donors in transamination. One of the nitrogens of urea is provided by glutamate.

This amino acid is also one of the three constituent acids of glutathione:

$$^{-}OOC-\underset{H}{\overset{\overset{+}{NH_3}}{C}}-(CH_2)_2-\overset{\overset{O}{\|}}{C}-\overset{H}{N}-Cys-\overset{\overset{O}{\|}}{C}-Gly$$

A product of glutamate, γ - aminobutyric acid (GABA), is essential for brain function becuase it retards transmission of nerve impulse. GABA is produced by the loss of the number one carboxyl group of glutamate:

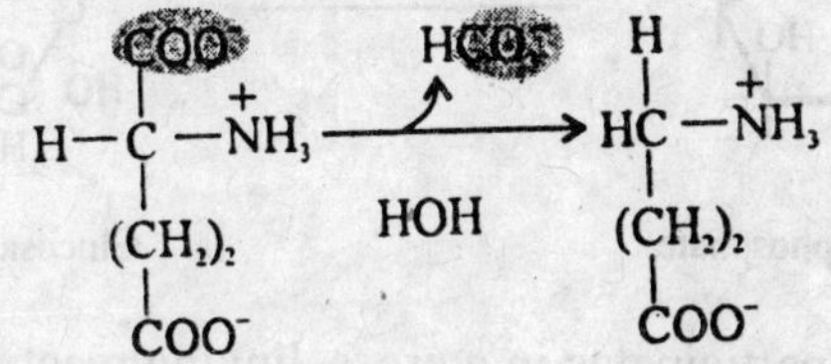

$$H-\overset{\displaystyle COO^-}{\underset{\displaystyle \underset{\displaystyle COO^-}{(CH_2)_2}}{C}}-\overset{+}{N}H_3 \xrightarrow[HOH]{HCO_3^-} H\overset{\displaystyle H}{\underset{\displaystyle \underset{\displaystyle COO^-}{(CH_2)_2}}{C}}-\overset{+}{N}H_3$$

It may be noted that 75 percent of the total free amino acids of brain can be accounted for by aspartate, glutamate, and their derivatives.

Amide Group of Glutamine as a Nitrogen Source in Biosynthetic Reactions. When converted to glutamine, the amide group on C–5 of glutamate becomes a prime source of nitrogen in many biosynthetic reaction. N^3 and N^9 of the purine ring system and the amino nitrogen at position 2 of guanylic acid are derived from glutamine:

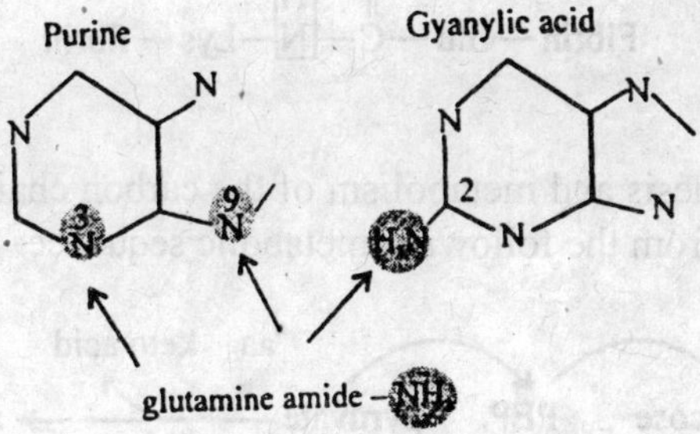

Purine Gyanylic acid

N N 3N 9N N N 2 H_2N N N

glutamine amide –NH_2

In the cytoplasm of eukaryotic cells is a carbamoyl phosphate synthetase that needs glutamine as a nitrogen donor. This carbamoyl phosphate synthetase provides a carbamoyl group for pyrimidine synthesis.

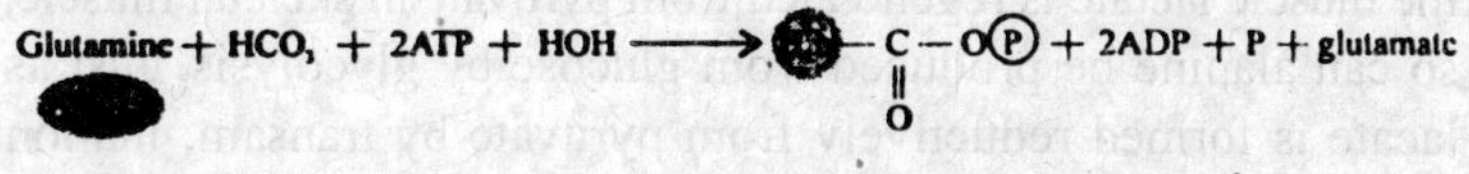

$$\text{Glutamine} + HCO_3 + 2ATP + HOH \longrightarrow NH_2-\underset{\displaystyle O}{\underset{\|}{C}}-O\text{Ⓟ} + 2ADP + P + \text{glutamate}$$

Glutamine also donates its amide nitrogen in the synthesis of amino sugars:

Fructose-6-phosphate —glutamine (-C(=O)-NH$_2$)→ Glucosamine-6-phosphate

Glutamine participates in a cross–linking reaction of fibrin in blood clotting process, resulting in the final insoluble polymeric form, or HARD CLOT:

Fibrin — glutamine — C(=O) — NH$_2$ — HN$^+$H$_2$ (F) — lysine — fibrin

↓ transamidase (→ NH$_3$)

Fibrin — Glu — C(=O) — NH — Lys — fibrin

Alanine

The synthesis and metabolism of the carbon chain iof alanine is apparent from the following metabolic sequence:

Glucose ⇄ PEP ⇄ pyruvate —(aa → keto acid)→ alanine

Alanine–Glucose Cycle. The Cori cycle represents a metabolic link between liver and muscle. The liver furnishes glucose, produced from lactate by gluconeogenesis, to skeletal muscle. In the muscle lactate is regenerated from pyruvate in skeletal muscle, so can alanine be produced from glucose by glycolysis. Just as lacate is formed reductively from pyruvate by transam, ination from a number of amino acids. Because of its interconvertibility with pyruvate by transamination, alanine can serve in the same manner as lactate in a recycling mechanism between skeletal muscle and the liver :–

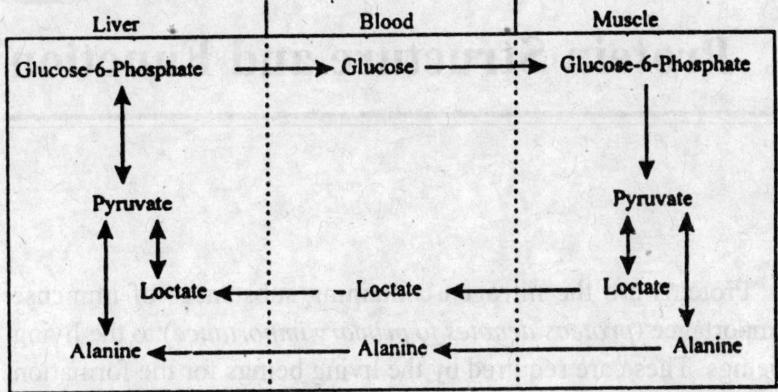

4

Protein Structure and Function

Proteins are the nitrogen containing substances of immense importance (*proteos denotes fo primary importance*) to the living beings. These are required by the living beings for the formation of **body tissues, enzymes, hormones, immunoglobulins** and other **intracellular** and **extracellular proteins.** Majority of the proteins are high molecular weight substances having very complex structures. On hydrolysis, proteins give **amino acids,** and some times certain non-protein residue(s) in addition to the amino

Figure 4.1 : ***Photomicrograph of a crystal of hexokinase, a key enzyme in the utilization of glucose.***

acids. Usually twenty different amino acids are obtained at the most in the protein hydrolysates. Thus, amino acids are the 'building stones' of the protein molecules. Amino acids in the different protein molecules remain arranged in different but definite sequences which impart a specific type of property in these molecules.

Biological functions of proteins

Proteins play crucial roles in virtually all biological processes. The significance and remarkable scope of their functions are exemplified in:

1. Enzymatic catalysis. Nearly all chemical reactions in biological systems are catalyzed by specific macromolecules called enzymes. Some of these reactions, such as the hydration of carbon dioxide, are quite simple. Others, such as the replication of an entire chromosome, are highly intricate. Nearly all enzymes exhibit enormous catalytic power. They usually enhance reaction rates by at least a millionfold. Indeed, chemical transformations rarely occur at perceptible rates *in vivo* in the absence of enzymes. Several thousand enzymes have been characterized, and many of them have been crystallized. The striking fact is that all known enzymes are proteins. Thus, poteins play the unique role of determining the pattern of chemical transformation in biological systems.

2. Transport and storage. Many small molecules and ions are transported by specific proteins. For example, hemoglobin transports oxygen in erythrocytes, whereas myoglobin, a related protein, transports oxygen in muscle. Iron is carried in the plasma of blood by transferrin and is stored in the liver as a complex with ferritin, a different protein.

3. Coordination motion. Proteins are the major component of muscle. Muscle contraction is accomplished by the sliding motion of two kinds of protein filaments. On the microscopic scale, such coordinated motions as the movement of chromosomes in mitosis and the propulsion of sperm by their flagella also are produced by contractile assemblies consisting of proteins.

4. Mechanical support. The high tensile strength of skin and bone is due to the presence of collagen, a fibrous protein.

5. Immune protection. Antibodies are highly specific proteins that recognize and combine with such foreign substances as viruses, bacteria, and cells from other organisms. Proteins thus play a vital role in distinguishing between self and nonself.

6. Generation and transmission of nerve impulses. The response of nerve cells to specific stimuli is mediated by receptor proteins. For example, rhodopsin is the photoreceptor protein in retinal rod cells. Receptor molecules that can be triggered by specific small molecules, such as acetylchaline, are responsible for transmitting nerve impulses at synapses—that is, at junctions between nerve cells.

7. Control of growth and differentiation. Controlled sequential expression of genetic information is essential for the orderly growth and differentiation of cells. Only a small fraction of the genome of a cell is expressed at any one time. In bacteria, repressor proteins are important control elements that silence specific segments of the DNA of a cell. A quite different way in which proteins act in different entiation is exemplified by nerve growth factor, a protein complex that guides the formation of neural networks in higher organisms.

CLASSIFICATION OF PROTEINS

A satisfactory classification of the proteins has not been possible till now as the chemical structure of most of the proteins is not known at present. The following classification is based merely upon solubilities and other physical properties of the proteins. On this basis, the proteins, have been classifed in three groups, namely **simple proteins, conjugated proteins** and **derived proteins.**

According to structure

Fibrous Proteins. These proteins are associated with cellular elements and often serve the function of supporting specific structures of the cell. Examples are wool, silk fibroin, collagen (connective tissue), myosin (muscle), keratin (hair) and fibrin (blood clot). These proteins are largely insoluble in aqueous media and have high molecular weights that cannot be accurately estimated because of difficulties encountered in their purification. They appear as fibers made up of linear molecules that are arranged roughly parallel to the fiber axis. They are amorphous and some are capale of stretching and contracting. Human fibrin has the molecular dimensions of 38 × 700 Å.

Globular Proteins. In contrast to fibrous proteins, the globular proteins are soluble in aqueous media and can be isolated in the crystalline state. Although not necessarily spherical, they are less asymmetric than fibrous proteins and consist of polypeptides (chains of amino acid residues) that are held together by cross-linked groups or in an aggregated state. Such aggregates are held together in a three dimensional structure by relatively weak non-covalent bonds.

ACCORDING TO SOLUBILITY

1. Albumins

These proteins are *soluble* in water and dilute salt solutions, and can be coagulated on heating. These can be precipitated from their solutions upon *fuel saturation* with ammonium sulphate. Albumins thus salted out may be purified by the process of dialysis. The coagulated albumins are insoluble in water, dilute acid, alkali and salt solutions. Coagulation starts above 75°C but this temperature differs for albumins obtained from different sources. The common examples of albumins are **ovalbumin, serum albumin, lactalbumin, myoalbumin.** The plant tissues have also been reported to contain albumins.

2. Globulins

These proteins are insoluble in water and soluble in dilute salt solutions, but can be coagulated on heating. These can be precipitated from aqueous solutions merely be fifty percent saturation with ammonium sulphate. These are found botn in the animal and plant tissues. The examples of globulins are **ovoglobulin** (*eggs*), **lactoglobulin** (*milk*), **fibrinogen, alpha, beta and gamma globulins** (*all in serum*), **myosin** and **tropmyosin** (*both in muscles*), **edestin** (*hempseed*) and **excelsin** (*Brazil nut*).

Globulins may be precipitated merely by diluting globulin solutions since these are soluble in dilute salt solutions but insoluble in salt-free water or extremely dilute salt solutions.

Globulins may also be precipitated merely by dialysis of the globulin solutions. Alubmins may be separated from globulins by using technique.

3.Globins

These are neutral type of histones which cannot be precipated by adding ammonium hydroxide as histones can be precipitated. Such proteins can easily combine with heme forming hemoglobin.

4. Glutelins.

These are insoluble in neutral aqueous solutions but soluble in dilute acid or alkali. Examples are glutenin (wheat) and oryzenin (rice).

5. Prolamines

Prolamines are *water insoluble* but *soluble* in 70 percent ethanol. These proteins are also insoluble in absolute alcohol and other neutral solvents. Such proteins are mainly found in the cereals. The examples of such proteins are **gliadin** of the wheat protein, zein (*corn protein*) and **hordein** (*barely protein*). Gliadin is found to be poor in lysine content. **Zein** and **hordein** are poor in lysine, tryptophan and valine content. The amide nitrogen content of prolamines is found to be very high.

6. Histones

More basic than most proteins, the histones tend to form complexes with acidic compounds in the cell (nucleic acids). They are soluble in water and insoluble in dilute ammonia solution. Examples are thymus histone (also called nucleohistone because it is found combined with nucleic acids and scombrone (mackerel.).

7. Protamines

Compared with most proteins, these are relatively small molecules. The protamines are soluble in water and basic in character. They are found associated with nucleic acids in the sperm of fish and are often called nucleoprotamines. Examples are salmine (salmon), sturine (sturgeon), clupeine (herring), and cyprinine(carp).

8. Albuminoids

These are characterized by their *marked insolubility* in water and in all other neutral solvents. The **keratins, elastins, collagens** and **keratohyaline** are all the examples of albuminoids. **Keratins** are present in the hairs, nails, horns, hoofs and feathers. The outermost layer of skin is also formed by keratin. **Keratohyaline** is present in the layer of skin. Keratohyaline is later on converted into keratin. These proteins are rich in arginine, lysine and histidine content. These can be hydrolysed only by concentrated acids and alkalies. Although these cannot be usually digested in the human alimentary canal but some of the moths (*cloth moths*) and micro organisms can digest such proteins. **Collagen** is found to exist in the connective tissues and in the bones, and are insoluble in all the neutral solvents. **Elastins** are found to be present in the yellow elastic fibers of the connective tissues. The various ligaments are purely elastins. These can be digested by **pepsin** and **trypsin.** The insolubility of the keratins has been very useful because it prevents dissolutions of body in water, or dilute acids and alkalies.

CONJUGATED PROTEINS

These are also known as compound proteins. Conjugated proteins are those proteins which contain in their molecular structure, in addition to protein, some non-protein organic substance(s) called prosthetic group or a metal. These proteins have also been subclassified into the following groups.

(a) Glycoproteins. Those proteins which contain carbohydrate in their molecular structure are included in this group. These include **mucopolysaccharides** and **mucoproteins.** The mucopolysaccharides contain predominantly carbohydrate and less protein, whereas the mucoproteins contain more protein and less carbohydrate. The **mucoproteins** include **ceruloplasmin** which is a blue copper containing protein, transferrin which is a iron transferring protein, **ossemucoprotein** of bones tendomucoproteins which is a iron transferring protein, osseomucoprotein of bones, **tendomucoproteins** which are found to be present in the tendons, **chonro-mucoproteins** which are found to be present in the cartilage, and **mucin** which is found in the digestive tract.

(b) Lipoproteins. These are those proteins which contain in their molecular structure, in addition to protein, a **lipid substance.** These are found to be soluble in the lipid-solvents. There exists a loose combination between protein and lipid. Such type of proteins occur mainly in the plasma, brain and eggs. In the plasma, these remain bound to mainly globulins. If such combinations contain α -globulin, these are known as the α -lipoproteins, and if β-globulins, then these are called β-lipoproteins.

(c) Nucleoproteins. In addition to nucleohistones and nucleoprotamines, there is a third group of proteins capable of binding nucleic acids. These nucleoproteins are not basic and hold the nucleic acid by secondary valence bonds. They occur in microorganisms and are soluble in isotonic salt solution.

(d) Chromoproteins. These proteins are pigmented owing to the prosthetic nonprotein group. They include hemoglobin (containing heme, and iron-protoporphyrin), ceruloplasmin (copper), hemocyanin (copper), ascorbic acid oxidase (copper) and ferritin (iron).

(e) Dehydrogenases. These are conjugated protein enzymes that possess oxidizing activity because of the presence of prosthetic groups such as NAD^+ (nicotinamide adenine dinucleotide), FMN (flavin mononucleotide), and FAD (flavin adenine dinucleotide).

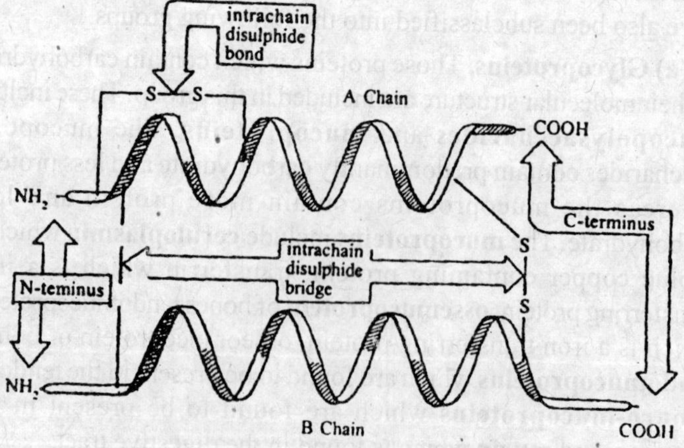

Figure. 4.2: ***Molecular structure of insulin.***

(f) Phosphoproteins. These proteins contain **phosphoric acid** in their molecular structure in addition to amino acids. Nucleoproteins and lipoproteins which do also contain phosphoric acid are not included in this group. The important examples of such proteins are **casein** and **vitellin.** Casein is found to be present in the milk, and vitellin is present in the egg-yolk.

(g) Metalloproteins. Such proteins contain **metal** (s) alone which remains attached to the protein. The examples of such protein are **ferritin** (metal portion iron), and **carbonic anhydrase** (Zn).

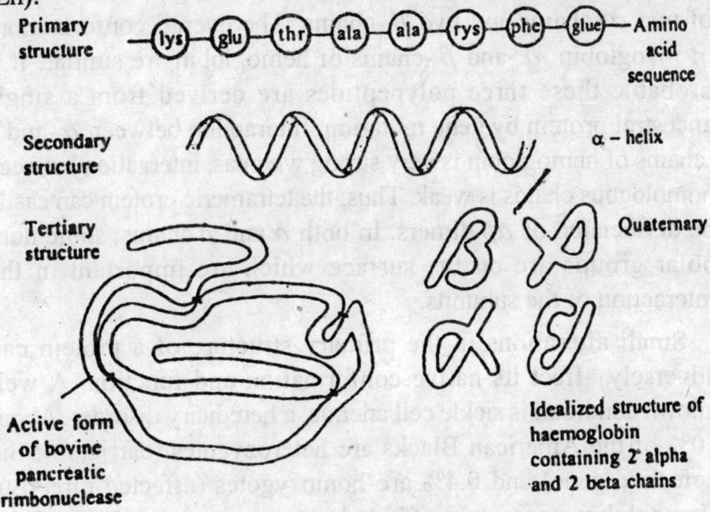

Figure. 4.3: ***The four levels of structural organization in protein molecules.***

Myoglobin and hemoglobin

Kendrew in 1950 studied the native conformation of myoglobin, the first globular protein the native conformation of myoglobin, the first globular protein to be studied by the X-ray crystallographic technique. Myoglobin, a heme protein from muscle is similar to hemoglobin in structure and function. Perutz studied the three dimensional structure of hemoglobin. Myoglobin, a single polypeptide with 153 amino acid residues has overall dimensions of 45Å + 30Å + 30Å. About 78% of the structure is in α -helical form and there are no β -pleats. In the case of hemoglobin also more than 70% of the structure is α-helix. This is not however, a characteristic feature of all globular proteins. α-Chymotrypsin, a proteolytic enzyme has relatively high proportion of β-pleats (45%) and low amount of α-helix and 12% β-pleats. Myoglobin has eight stretches of α-helix interspersed with non-ordered segments. The heme group is located in a crevice between two helices. The prosthetic group is on the surface facilitating oxygen binding. Most of the hydrophobic side chains are buried inside and the polar residues are on the surface. As indicated earlier, hemoglobin is made up of two α-chains and two β-chains. The overall conformations of myoglobin, α-and β-chains of hemoglobin are similar. It is probable these three polypeptides are derived from a single ancestral protein by gene mutation. Interaction between α-and β-chains of hemoglobin is very strong whereas, interaction between homologous chains is weak. Thus, the tetrameric protein can easily be dissociated of $\alpha\beta$ dimers. In both α and β chains, some non-polar groups are on the surface which are important in the interaction of the subunits.

Small alterations in the primary structure of a protein can adversely affect its native conformation and function. A well known condition is sickle cell anemia, a hereditary disorder. About 10% of the American Blacks are heterozygotes (carriers of the defective gene) and 0.4% are homozygotes (affected directly). Hemoglobin content in affected persons is about 50% of the normal value, since they suffer from repeated crisis brought about by physical exercise. They become weak and dizzy and are short of breath. Microscopic examination will show that erythrocytes are abnormal in shape. The cells are long and crescent-like akin to the blade of a sickle. The only alteration in the primary structure is the substitution of valine in place of glutamate in the sixth position in the β-chains. This results in what is known as a 'sticky path' which causes aggregation of hemoglobin molecules when O_2 content is low.

Protein Denaturation

Since the conformation of a protein is solely dependent on week valence forces it can be disrupted by a variety of physical and chemical agents. The process is known as denaturation. The organized structure is lost and the protein assumes a highly disordered form. The protein will lose its biological activity. It becomes more susceptible to proteolytic attack. Many globular proteins are rendered less soluble on denaturation. An interesting instance of a protein losing its biological activity on denaturation is with reference to monellin, a protein 300 times sweeter than sucrose. Monellin isolated from an African plant *Dioscorephyllum cumminsis* is made up of two polypeptide chains made up of 44 and 55 amino acid residues, non-covalently associated. On denaturation the protein loses its sweet taste. If the denaturation process is not very drastic the process becomes reversible, once the denaturing agent is removed. This phase is known as renaturation.

Moist heat is powerful denaturing agent. Hydrogen bonds are easily disrupted by heat. Extremes of acidity and alkalinity at moderate temperature, disrupt electrostatic interactions and cause conformational changes. Heavy metal ions like Hg^{++} and Pb^{++} can complex with negative charges or sulphydryl groups. The basis of administering egg white immediately to counteract lead poisoning is based on the complexation of egg proteins with the toxic metal ion. Polar organic solvents like ethanol also can cause denaturation, if temperature is not controlled. Trichloroacetic acid and perchloric acid are two common laboratory reagents used for denaturation and precipitation of proteins. Sodium dodecyl sulphate (SDS, CH_3 $(CH_2)_{10}$ $CJ_2 - OSO_3^-$ Na^+) another reagent used in protein analysis is also a powerful denaturing agent. It can dissociate subunits by disrupting non-ploar interactions. Urea ($H_2N-CO-NH_2$) and guanidine hydrochloride [$H_2N - C\ (NH_2) = N^+\ H_2Cl^-$] at high concentrations (6.0 – 8.0 M and 2.0 – 3.0 M respectively) act as denaturing agents. Unlike commonly believed, urea does not act by disrupting hydrogen bonding. It acts by interfering with non-polar interactions.

Plasma Proteins

Hemoglobin the major protein in blood (12-15 g/100 ml) is

confined to the red cells. Plasma, the cell-free fraction is also rich in proteins (6.3-7.5 g/100 ml). A large array of structurally and functionally different proteins is present in plasma. There are carrier proteins like albumin, transferrin and apolipoproteins. Quantitatively the second important group of proteins after albumin are the immunoglobulins, which are defence proteins. Another group of proteins known as protein inhibitors accounts for 10% of total plasma proteins. A variety of clotting factors which are mostly precursors for proteolytic enzymes involved in blood coagulation are present in relatively low concentrations in plasma. There are circulating protein hormones like insulin and thyroid stimulating hormome in very minuter amounts. Enzymes which are mostly nonfunctional in plasma are also present in traces.

Several techniques are employed for the qualitative and quantitative separation of plasma proteins. A common technique used in clinical laboratories for the separation of serum (rather than plasma to eliminate fibrinogen) is zone electrophoresis on paper, agar gel or cellulose acetate strips at pH 8.6. Five fractions in the order of increasing mobility towards anode namely albumin, α -1, α -2, β and – γ globulins are usually obtained by this technique. Each globulin fraction is comprised of more than one protein which can further be separtated by sophisticated techniques.

A. After the electrophoretic run the protein bands are stained to visualize the proteins with suitable dyes. Proteins bind tightly to coloured dyes through electrostatic and hydrophobic interactions.

B. Semiquantitation of the protein bands by scanning in a densitometer.

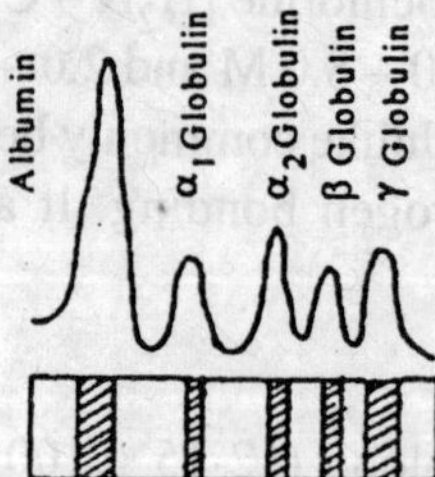

Figure. 4.4: *Electrophoretic pattern of serum proteins*

TABLE 4.1. PLASMA PROTEINS

Proteins	Concentration	Mr	Function
G/100 ml	kDa		
Prealbumin	0.03	61	Transport of thyroxine
Albumin	3.2-4.2	68	Diverse functions
α_1 -Acid glycoprotein	0.10	44	Acute phase protein
α_1 -Proteinase inhibitor	0.25	54	Elastase inhibitor
α_2-Macroglobulin	0.20	720	Nonspecific proteinase
inhibitor			
Haptoglobin	0.10	85	Binds hemoglobin
Ceruloplasmin	0.02	150	Copper transport
Prothrombin	0.02	63	Blood clotting
Transferrin	0.30	85	Transport of iron
Plasminogen	0.05	140	Retraction of fibrin clot
Fibrinogen	0.30	340	Precursor of fibrin clot
γ-Globulins	0.8-1.7	–	Immune reaction

Albumin the major plasma protein, accounts for about 55-60% of the total proteins. It is synthesized exclusively in liver and a healthy liver can synthesize 10-15 g of albumin per day. The life span of an albumin molecule is about 20-25 days. Nearly 40% of plasma calcium is bound to albumin. Albumin is involved in the transport of free fatty acids from adipocyte to liver. Bilirubin, steroid hormones and many water insoluble drugs are also bound and transported by albumin. Along with other plasma proteins albumin acts as a buffer. Albumin has a major role in osmotic regulation and fluid distribution. It accounts for about 80% of the total colloidal osmotic pressure (25 mm Hg) in blood plasma. One g of albumin can hold 18.0 ml of water. Decreases in albumin level will cause fluid accumulation in the interstitial space and soft tissues resulting in edema. Prolonged malnutrition, kidney diseases associated with protein loss in urine, liver cirrhosis and extravagation as in burns will cause this condition. However, in analbuminemia a rare genetic disorder, persons do not develop edema and are apparently healthy. In this condition globumin level increases and arterial blood pressure is lowered to compensate for the absence of albumin.

The predominant component of α_1 -globulin (0.3-0.5 g/100 ml) is a glycoprotein known as α_1 -proteinase inhibitor which is a powerful inhibitor of neutrophil elastase. Elastase when not regulated can cause tissue breakdown and loss of elasticity in the lung. Two variants of this protein (S and Z) which are far less effective against elastase are known mainly in the Scandinavian countries. Persons with these modified α_1 -proteinase inhibitors develop emphysema α_1 -Acid glycoprotein is so called because of its low isoelectric point. The concentration of this protein is significantly increased in plasma in inflammation and other acute disease conditions. Nearly 30% of the α_2 -fraction and 40% of β -fraction are accounted for by apolipoproteins, which are described later. The functions of transferrin and haptoglobin which belong to the β and α_2 -globulin fractions are also detailed later. Ceruloplasmin a copper containing protein belongs to α_2 - macroglobulin which has the capacity to trap and remove from circulation a large number of proteinases and can hence play a regulatory role.

γ-Globulins

This group of proteins which are unique to vertebrates are also called immunoglobulins. Unlike most other plasma proteins which are synthesized in liver, immunoglobulins are made β - lymphocytes and plasma cells of lymph nodes, bone marrow and spleen. They bind specifically to antigenic sites on other molecules. The antigens, which provide the binding sites, can be proteins, polysaccharides nucleic acids or even relatively low molecular weight compounds, which are of foreign origin to the body. Thus, bacterial and viral components can act as antigens and trigger antigen-antibody reactions. Models of such reactions are shown in the figure.

Knowledge of immunoglobulins came from the studies on protein in plasma and urine in patients suffering from myelomatosis (immunocytoma tumors). In this condition, γ - globulin level is increased in serum. Characteristic proteins known as Bence-Jones proteins are excreted in large amounts in urine. These proteins precipitate on heating urine samples to 45-50°C but redissolve at higher temperatures. They were identified as low molecular weight proteins (light chains) which are constituent units of immunoglobulins. Two types of light chains known as κ

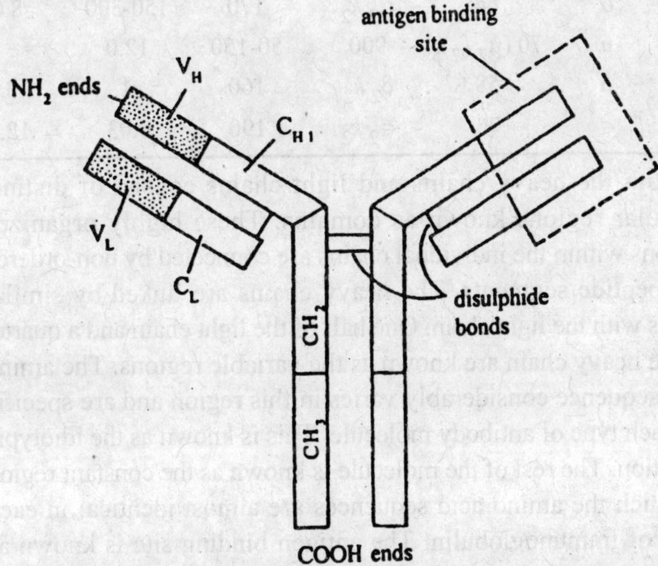

Figure. 4.5: ***Molecular structure of an immunoglobulin molecule.***

and λ chains are present in immunoglobulins and in humans the former predominates. Each immunoglobulin molecule is a heterotetramer made up of two light chains and two heavy chains. Light chains are simple proteins whereas, the heavy chains are glycoproteins. The light and heavy chains are synthesized as separate molecules and are assembled to funtional immunoglobulins in the cells of their origin. The immunoglobulins are classified into five groups based on the type of heavy chain.

In general, Ig molecules can be represent as Y -shaped structure. Ig M is a pentameric structure. A polypeptide known as J-chain (Mr 15 kDa) links the monomeric units through disulphide bridges. Ig A also has J- chain and in this instance, aggregation stops at the dimer stage.

TABLE 4.2: CLASSIFICATION OF IMMUNOGLOBULINS (Ig)

Group	Heavy Chain kDa	Mr form	Native Mr, kDa	Total concentration	Plasma carbohydrate mg/100 ml	Percent
Ig G	γ	53	$\gamma_2 k_2$	150	800-1200	3.0
Ig A	α	64	$\alpha_2 k_2$	170	150-300	8.0
Ig M	μ	70 $(\mu_2 k_2)_5$	900	50-150	12.0	
Ig D	δ	58	$\delta_2 k_2$	160	4	13.0
Ig E	ϵ	75	$\epsilon_2 k_2$	190	0.03	12.0

Both the heavy chains and light chains consist of distinct globular regions known as domains. These highly organized regions within the individual chains are connected by non-ordered polypeptide segments. The heavy chains are linked by similar bonds with the light chain. One half of the light chain and a quarter of the heavy chain are known as the variable regions. The amino acid sequence considerably varies in this region and are specific for each type of antibody molecule. This is known as the idiotypic variation. The rest of the molecule is known as the constant region in which the amino acid sequences are almost identical in each class of immunoglobulin. The antigen binding site is known as Fab site consisting of light chains and the N-terminal half of the heavy chain. The remaining part of immunoglobulin known as Fc also has important biological functions. For example, the Fc region of immunoglobulin G is responsible for triggering pathways of the immune response that leads to the lysis of unwanted organisms. An example is the complement system that is activated which consists of a series of proteinases.

Ig G is mainly responsible for humoral immunity. It is the only immunoglobulin that can cross through the placenta. Ig A which is present in some secretions is thought to provide surface immunity. Ig A secretion is facilitated by its specific binding to a protein known as secretory component. Ig M is the receptor in the B-lymphocytes and helps in established humoral immunity. Ig D is also found at the surface of B-lymphocytes but its function is not known. Ig E is found in mast cells and basophils. The concentration of Ig E increases in response to allergic reactions.

Disorders of immunoglobulins are numerous. Agammaglobulinemia is a rare X-chromosome associated genetic disease affecting only the males. In this condition, there is virtual absence of immunoglobulins in blood plasma. The affected persons are highly susceptible to bacterial infection but react almost normally to viral infections. Hypogammaglobulinemia may be restricted to a single class of immunoglobulins or may involve under-production of all the five types. In myelomas, increased production of a restricted class of immunoglobulins or of a single specific immunoglobulin is seen. Immunoglobulin level in blood is increased generally in infections. Sometimes the body rejects its own proteins which become antigenic. This results in the autoimmune disorders. Systemic lupus erythomatosis and some types of rheumatoid arthritis are examples. It is probable, buried antigenic sites (epitopes) in endogenous proteins get expose or the 'self' proteins bind with exogenous triggers and become antigenic in these conditions.

Primary Structure

One of the great achievements of modern chemistry has been the development of techniques for determining the precise sequence of amino acids in a protein. The first protein to have its sequence of amino acids determined was the hormone insulin. Its molecular formula is $C_{254}H_{377}N_{65}O_{75}S_6$. This is small as proteins go. Even so, it took Dr. Frederick Sanger and his colleagues at the University of Cambridge in England 10 years (1944-1954) to work out the exact sequence of amino acids in the protein. Insulin consists of two polypeptide chains containing a total of 51 amino acid residues.

The amino acid cysteine occurs at six positions in the insulin molecules. Wherever two cysteines are close to each other, they can be oxidized (each losing a hydrogen atom). As a result a covalent bond forms between their respective sulfur atoms forming a **disulfide bridge**. In this way two different polypeptides can be drawn into a loop. Sanger found one *intra* chain and two *inter* chain disulfide bridges in the insulin molecule.

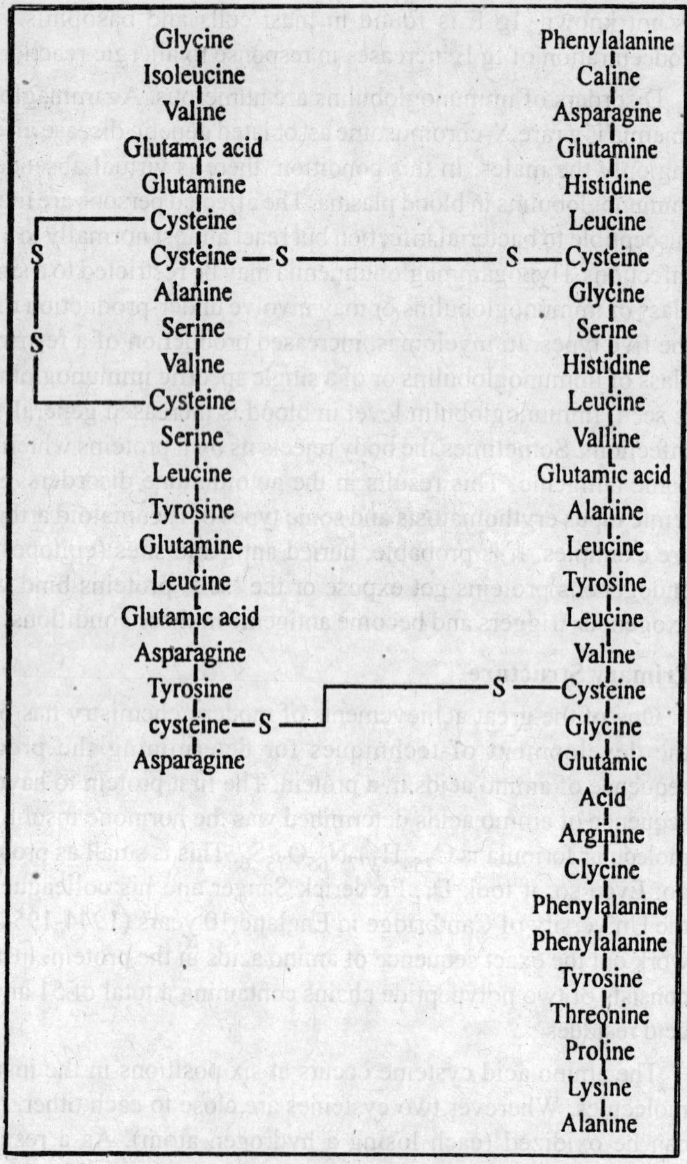

Figure. 4.6: ***Sequence of amino acids in the human insulin molecule. The molecule consists of two polypeptide chains held together by two disulfide bridges.***

Since the days of Sanger's pioneering work, the techniques of protein sequencing have developed rapidly. It is now possible to have much of the job done automatically by machines. The sequence of amino acids is now known for hundreds of different proteins. The sequence of amino acids in a protein, together with the location of any disulfide bridges, is called the **primary structure** of the protein.

Let us now examine the primary structure of **lysozyme.** Lysozyme is an enzyme found in an egg white, tears, and other secretions. It is responsible for breaking down the polysaccharide walls of many kinds of bacteria, and thus it provides a measure of protection against infection.

The lysozyme in an egg white is a simple polypeptide containing 129 amino acid residues. There are four pairs of cysteines, establishing disulfide (S-S) bridges between positions 6 and 127, 30 and 115, 64 and 80, and 76 and 94. The presence of these covalent linkages tells us immediately that we cannot represent the polypeptide as a straight, rigid chain. The chain must fold on itself to allow the pairs of cysteines to be close to each other.

Secondary Structure

To discover the actual configuration of a protein molecules in three dimensional space, we must turn to another analytical technique, that of x-ray crystallography. A crystal consists of an orderly, stacked arrow of ions or molecules. If a beam of x-rays is passed through a crystal, some of the x-rays will be reflected by atoms in the crystal. Because of the orderly arrangement of the atoms, the reflections will also be orderly. A simple analogy would be the orderly pattern produced by a street light at night when viewed through the grid of a window screen. By examining the pattern of reflections produced as the beam is directed at the crystal from a variety of different angles, it is possible (especially with the help of a computer) to determine the arrangement of atoms in that crystal.

The first proteins to be analyzed by x-ray crystallography were certain structural proteins like the alpha-keratins of wool and fibroin, the protein secreted by the silkworm. These proteins were good choices for study because of their regular, self-repeating structures.

X-ray analysis of alpha-keratins revealed several interesting facts.

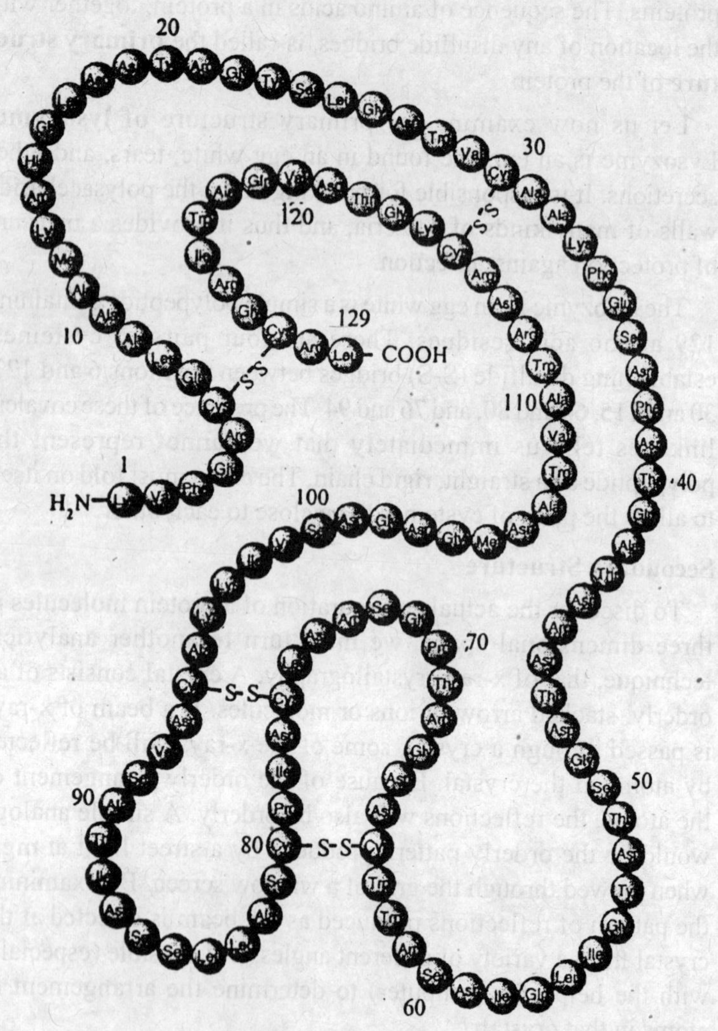

Figure. 4.7: ***Primary structure of lysozyme. The molecule consists of a single polypeptide chain of 129 amino acid residues. They are numbered beginning at the amino terminal. There are four intrachain disulfide bridges.***

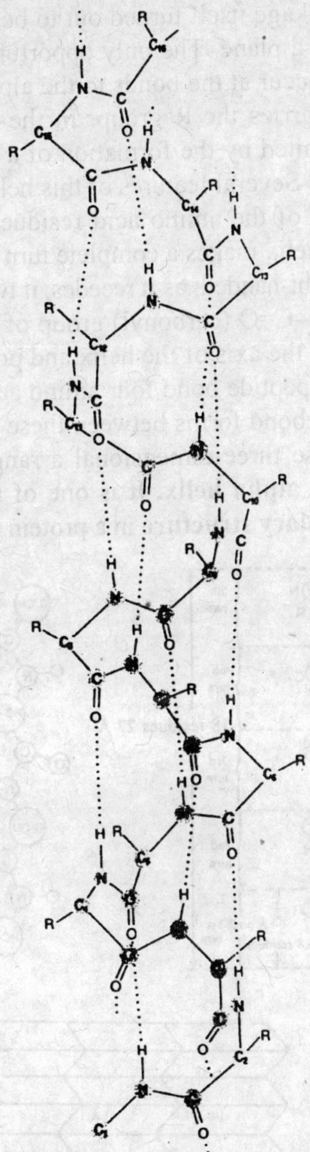

Figure. 4.8: ***The alpha helix, a common form of secondary structure turn every 3.6 residues. Note the hydrogen bond that forms between the – C = O group of each peptide bond and the –N–H group of the fourth peptide bond below it in the helix.***

The peptide linkage itself turned out to be very rigid, with all its atoms lying in a plane. The only opportunities for flexibility in a polypeptide occur at the bonds to the alpha carbon—that is, the carbon that carries the R group. In the alpha-keratins this flexibility is exploited by the formation of a helical twist to the polypeptide chain. Several features of this helix should be noted. (1) The R groups of the amino acid residues all extend to the outsides. (2) The helix makes a complete turn every 3.6 residues. (3) The helix is right-handed; as it recedes, it twists in a clockwise direction. (4) The –C=O (carbonyl) group of each peptide bond extends parallel to the axis of the helix and points directly at the –NH group of the peptide bond four amino acids below it in the helix. A hydrogen bond forms between these groups: –C=O . . . H–N–. This precise three-dimensional arrangements of amino acids is called the **alpha helix.** It is one of the most common examples of **secondary structure** in a protein.

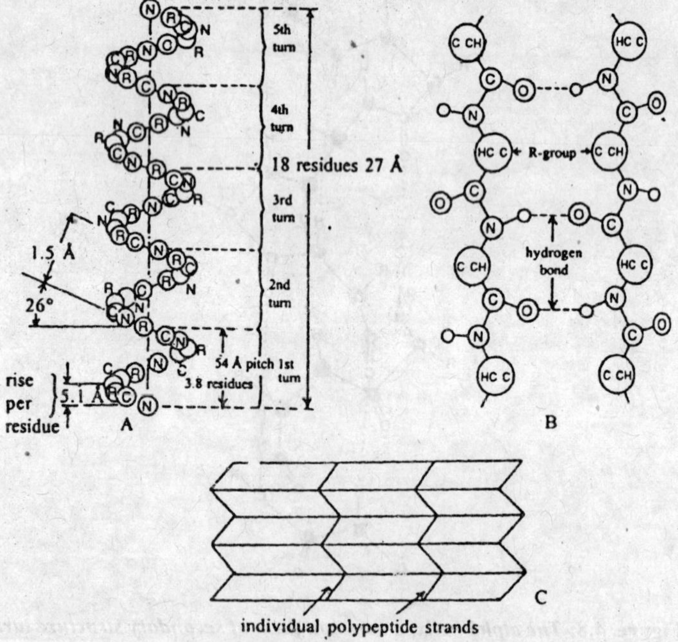

Figure. 4.9: ***A-C- Secondary structure of proteins: A — arrangement of polypeptide chains in (-helix; B-C— β -pleated sheets; B—extended β - (antiparallel) plypeptide strands; C—schematic of pleats (top view).***

What makes the alpha helix possible is the large number of hydrogen bonds. The strength of a single hydrogen bond is only about 5% of that of a covalent bond. But the presence of a hydrogen bond between each amino acid in the chain, and the one that is four amino acids away, provides sufficient total bond strength to make the alpha helix a very stable structure. However, it should be possible by breaking these intrachain hydrogen bonds to pull the helix out in much the same way that a spring can be extended. And, in fact, wool fibers can be stretched to about twice their normal length thanks to the molecular properties of the alpha helix.

Fibroin is the protein that silkworms use to spin the threads of their cocoon. It, too, has been subjected to x-ray analysis. In fibroin the polypeptide chains are extended. A number of chains lie parallel to each other, held together by the hydrogen bonds that form between the –C - O and –N–H groups of one chain with the –N–H and – C - O groups of the adjacent chain. In silk it would be better to describe the chains as being antiparallel because adjacent chains run in opposite directions—that is, from N-terminal to C-terminal and vice versa. A series of antiparallel stands lying side by side make up a beta-pleated sheet. It is another important and commonly found example of secondary structure. Because the chains are fully extended, we would not expect to be able to stretch them without breaking covalent bonds. It is for this reason that silk is not extensible. However, the layers of beta-pleated sheet in silk fibers make them very supple.

The alpha helix and the beta-pleated sheet are two of the most common examples of secondary structure, but other orderly arrangements of amino acid residues also occur in polypeptides. All examples of secondary structure are a direct reflection of the particular amino acids and their order that are used in the synthesis of the polypeptide. The chains of fibroin, for example, consist mainly of alternating glycine (hydrophobic) residues. This arrangement enables the chains to rest closes together in the beta-pleated sheet configuration. On the other hand, a chain containing a series of alanine residues will spontaneously fold into an alpha helix with the CH_3 groups of the alanines projecting to the exterior of the helix

Figure. 4.10: ***Collagen triple-helix structure.***

Stages, as with cellulose. The fact that the protein is extremely resistant to stretching is an essential part of its functioning. For example in tendons, bone, skin, teeth, and connective tissue. Proteins which entirely in the form of helical coils, such as keratin and collagen, are exceptional.

Tertiary Structure

Usually the polypeptide chain bends and folds extensively, forming a precise,compact 'globular' shape. This is the protein's tertiary structure and it is maintained by the interaction of the four types of bond already discussed, namely ionic, hydrogen and disulphide bonds as well as hydrophobic interactions. The latter are quantitatively the most important and occur when the protein folds so as to shield hydrophobic side groups from the aqueous surroundings, at the same time exposing hydrophilic side chains, as described above.

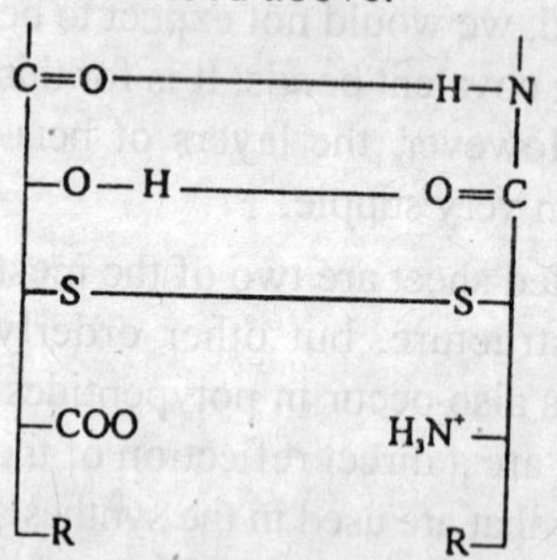

Hydrogen bond between CO and NH group of main amino acid chain

Hydrogen bond between R groups

Disulphide bond formed between cysteine residues

Ionic bond between charged R groups

Hydrophobic interaction of non-popular R groups

Figure 4.11: ***Summary of types of bonds stabilising secondary and tertiary structures of proteins. Hydrophobic interactions (associations of non-polar molecules or parts of molecules) to exclude water molecules in the aqueous environment of the cell are particularly important in maintaining structure, as in membranes.***

The tertiary structure of a protein can be determined by X-ray crystallography. By early 1959, and after many year's work, John kendrew and Max Perutz had built the first atomic model of myoglobin showing secondary and tertiary structures using this technique. They received the Nobel Prize for their work in 1962:

Primary structure – single polypeptide chain of 153 amino acids, the sequence was elucidated in the early 1960s;

Secondary structure – about 75% of the chain is α -helical (8 helical sections).

H_2N—Val—Leu—Ser—Glu—Gly—Glu—Trp—Gin—Leu—Val—Leu—His—Val—Tyr—Ala—Lys—Val—
Glu—Ala—Asp—Val—Ala—Gly—His—Gly—Gin—Asp—Ile—Leu—Ile—Arg—Leu—Phe—Lys—
Ser—His—Pro—Glu—Thr—Leu—Glu—Lys—Phe—Asp—Arg—Phe—Lys—His—Leu—Lys—Thr
Glu—Ala—Glu—Met—Lys—Ala—Ser—Glu—Asp—Leu—Lys—Gly—His—His—Glu—Ala—Glu—
Leu—Thr—Ala—Leu—Gly—Ala—Ile—Leu—Lys—Lys—Gly—His—His—Glu—Ala—Glu—
Leu—Lys—Pro—Leu—Ala—Gin—Ser—His—Ala—Thr—Lys—His—Lys—Ile—Pro—Ile—Lys—
Tyr—Leu—Glu—Phe—Ile—Ser—Glu—Ala—Ile—Ile—His—Val—Leu—His—Ser—Arg—His—
Pro—Gly—Asn—Phe—Gly—Ala—Asp—Ala—Gin—Gly—Ala—Met—Asn—Lys—Ala—Leu—Glu—
Leu—Phe—Arg—Lys—Asp—Ile—Ala—Ala—Lys—Tyr—Lys—Glu—Leu—Gly—Tyr—Gin—Gly—COOH

Figure 4.12: ***Primary structure of myoglobin***

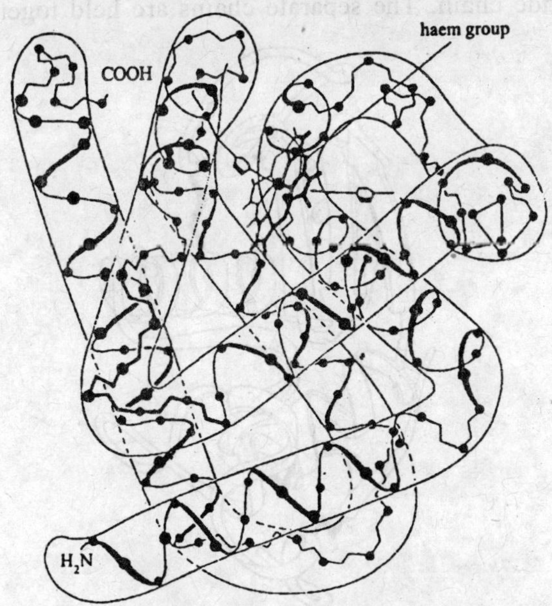

Figure 4.13: ***Conformation of myoglobin deduced from high resolution X-ray data. There are eight sections of α helix surrounding a haem group which forms a flat disc.***

Tertiary structure — non-uniform folding of the α -helical chain into a compact shape;

Prosthetic group — group (contains iron).

Myoglobin is formed in muscle where its function is to store oxygen. Oxygen combines with the haem group as in haemoglobin. The haem group gives its red appearance. The elucidation of tertiary structure is still very time-consuming. Use of computers and other techniques to predict tertiary structure, based on knowledge of primary and secondary structures, is a fast-growing area of molecular biology. From this follows the possibility of designing proteins with particular shapes for particular functions, with important applications is industry and medicine.

Quaternary Structure

Many highly complex proteins consist of more than one polypeptide chain. The separate chains are held together by

Figure 4.14 : ***Tertiary structure of lysozyme. The arrows represented a region of β sheet. α helices are shown as regular coils. The rest of the molecule is represented as ribbon and the four disulphide bridges as black zig-zags.***

hydrophobic interactions and hydrogen and ionic bonds. Their precise arrangement is known as the **quaternary structure.** Haemoglobin shows such a structure. It is the red oxygen-carrying pigment found in the red blood cells of vertebrates. It consists of four separate polypeptide chains of two types, namely two α chains and two β chains. These resemble myoglobin in structure. The two α-chains each contain 141 amino acids, while the two α chains each contain 146 amino acids. The complete structure of haemoglobin was worked out by Kendrew and Perutz.

As is typical of globular proteins, its hydrophobic side chains point inwards to the centre of the molecule, and its hydrophilic side chains face outwards, making it soluble in water. A mutation which causes one of the hydrophilic amino acids to be replaced by a hydrophobic amino acid, thereby reducing its solubility, is responsible for the disease sickle cell anaemia.

The protein coats of some viruses, such as the tobacco mosaic virus, are composed of many polypeptide chains arranged in a highly ordered fashion.

Figure. 4.15: ***Structure of haemolgobin. The molecule consists of four chains: two a chains and two β chains. Each chain carries α haem to which one molecule of oxygen binds. The assembly of a protein from separate polypeptide chains is an example of quaternary structure.***

Denaturation and renaturation of proteins

Denaturation is the loss of the specific three-dimensional shape of a protein molecule. The change may be temporary or permanent, but the amino acid sequence of the proten remains unaffected. If denaturation occurs, the molecule unfolds and can no longer perform its normal biological function. A number of agents may cause denaturation as follows.

Heat or radiation. eg infra-red or ultra-violet light. Kenetic energy is supplied to the protein causing its atoms to vibrate violently, so disrupting the weak hydrogen and ionic bonds. Coagulation of the protein then occurs.

Strong acids and alkalis and high concentrations of salts. Ionic bonds are disrupted and the protein is coagulated. Breakage of peptide bonds may occur if the protein is allowed to remain mixed with the reagent for a long period of time.

Heavy metals. The positively charged ions of heavy metals (cations) form strong bonds with the negatively charged carboxyl groups on the R groups of proteins and often disrupt ionic bonds. They also reduce the proteins and often disrupt ionic bonds. They also reduce the protein's electrical polarity (its overall charge) and thus increase its insolubility. This causes the protein to precipitate out of solution.

Organic solvents and detergents. These reagents disrupt hydrophobic interactions and form bonds with hydrophobic (non-polar) groups. This in turn causes the disruption of hydrogen bonding. When alcohol is used as a disinfectant it functions to denature the protein of any bacteria present.

Renaturation

Sometimes a protein will spontaneously refold into its original structure after denaturation, providing conditions are suitable. This is called renaturation, and is good evidence that tertiary structure can be detemined purely by primary structure and that biological structures can spontaneously assemble according to a few genera principles.

5

The Catalysts of Life

Enzymes are the biological catalysts and are proteinaceous in nature. They are able to accelerate the rate of chemical reactions in a living cell. Like catalysts, they are not used up in the reaction, but unlike catalysis they are produced by living cells only. These are the enzymes which run myriads of biochemical reactions under normal cellular conditions. For hydrolyzing starch in the food for example, high temperature or acidity is required. In the alimentary canal however, it is digested to glucose under relatively normal conditions of temperature and pH by the enzymes. Similarly the enzymes are able to hydrolyse other macromolecules such as the protein and lipid also under normal physiological conditions.

HISTORICAL

Earlier the agents causing fermentation of sugars were named as ferments. Pasture working on microorganisms concluded that fermentation and similar processes could be performed only by living cells. This concept was later disputed by J. Liebig who proposed that fermentation could take place even in the absence of living cells. Therefore, a distinction was made between organized ferments present in cells and unorganized ferments that were not associated with microorganism. In 1836, J. J. Berzelius named the biological agent as 'diastase' which is known today as amylase capable of converting starch from malt extract into sugar. The term enzyme ('in yeast') was proposed by Kühne in

1877 to distinguish enzyme from organized and unorganized ferments.

OCCURRENCE AND DISTRIBUTION

Enzymes occur in all living cells but not all enzymes are found in all the cells. The enzymes catalyze a wide variety of biochemical reactions many of which are localized in specific organs or are peculiar to certain species of plant or animal life. Thus, for example, pepsin is produced only in the cells of gastric mucosa and trypsin only in the pancreas. In the plant world lipases are not generally distributed but are found chiefly in plants that produce oilseeds.

A few of the enzymes are present in most forms of life. For example, catalases and peroxidases are widely distributed in all higher plants and animals.

The amount of enzymes may be different in different tissues. For example, resting seeds have low activity of analyses and proteinases while the germinated seedlings have more activity. In some fruits, enzymes are concentrated near pit. Some enzymes are also organelle specific.

NOMENCLATURE

Naming of enzymes

In the past, enzymes were named in a haphazard manner as and when they were discovered. The ways in which enzymes are or have been named are listed below:

1. The first enzymes studied have been named for their colour, their localization within the body or after the person who discovered them. However, this nomenclature had not been agreeable to many.
2. Later, many enzymes have been named by adding the suffix '-ase' to the name of the substrate, for example, urease catalyzes the hydrolysis of arginine to ornithine and urea and so on. However, this nomenclature has not been always practicable.
3. Further, many other enzymes have been given chemically

uninformative names, e.g., pepsin, trypsin and catalase. With the increasing number of newly discovered enzymes, this kind of nomenclature has not been accepted, this system has its limitations.

4. Finally, a systematic classification of enzymes has been adopted on the recommendation of an International Enzyme Commission as listed in the 1973 edition of Enzyme Nomenclature with a few exceptions. According to this, enzymes have been classified into six major classes and sets of subclasses based on the nature and type of reactions catalyzed. According to this, each enzymes is assigned:

A recommended name: It is usually short and appropriate for everyday use.

A systematic name: It identifies the reaction the enzyme catalyzes, and

A classification number: It is used where accurate and unambiguous identification of an enzyme is required, as in research journals, abstracts and indexes.

An example is given by the enzymes catalyzing the reaction:

$$\textbf{ATP + creatine} \rightleftharpoons \textbf{ADP + phosphocreatine}$$

The recommended name : creatine kinase

The systematic name : ATP : creatine phosphotransferase

Classification number : EC 2.7.3.2, where EC stands for Enzyme Commission, the first digit (2) for the class name (transferases), the second digit (7) for the subclass (phosphotransferases), the third digit (3) for the sub-subclass (phosphotransferases with a nitrogenous group as acceptor) and the fourth digit (2) designates creatine kinase.

Enzyme classification

Enzymes can be classified into six major classes based on the nature and type of reaction catalyzed as given below:

1. **Oxidoreductases :** These enzymes catalyze oxidation or reduction reactions by transfer of hydrogen or electrons. e.g., succinic deydrogenase.

$$\begin{matrix} CH_2-COOH \\ | \\ CH_2-COOH \end{matrix} + FAD \xrightleftharpoons{\text{succinic dehydrogenase}} \begin{matrix} H & & COOH \\ & C & \\ & \| & \\ & C & \\ HOOC & & H \end{matrix} + FADH$$

Succinic acid **Fumaric acid**

2. **Transferases.** The enzymes are involved in transferring functional groups between a donor and an acceptor molecule. For example, glutamate: pyruvate amino transferase transfers amino group from glutamate to pyruvate. Glutamate is deaminated to produce 2-oxoglutarate, and pyruvate is aminated to produce alanine;

$$\begin{matrix} COOH \\ | \\ H-C-NH_2 \\ | \\ CH_2 \\ | \\ CH_2 \\ | \\ COOH \end{matrix} + \underset{\text{pyruvate}}{CH_3CO.COOH} \rightleftharpoons \begin{matrix} COOH \\ | \\ C=O \\ | \\ CH_2 \\ | \\ CH_2 \\ | \\ COOH \end{matrix} + \underset{\text{alanine}}{CH_3CH(NH_2)COOH}$$

glutamate **2-oxoglutarate**

Important groups of enzymes in this class are:

(i) Aminotransferases or transaminases which transfer amino group

(ii) Kinases, which transfer phosphoryl group from ATP or another nucleoside triphosphate to alcohol or amino group acceptors.

(iii) Glycosyl transferases which catalyse the transfer of an activated glycosyl residue to a glycogen primer

(iv) Transaldolase and transketolase which catalyze the transfer of aldol or ketol groups.

3. Hydrolases. These enzymes are special kinds of tranferases in which the donor group is transferred to water molecule. In other words, the reaction is that of a hydrolytic cleavage. The bonds which are normally subjected to hydrolytic cleavage are C—O, C—N, O—P and C—S bonds. An example of hydrolase is a lipase which hydrolyzes a triacyl glycerol (fat) into fatty acids and glycerol;

$$\begin{array}{lcccccc} CH_2O.COR & & & & CH_2OH & & \\ | & & & & | & & \\ CHO.COR & + & 3\,H_2O & \rightarrow & CHOH & + & 3\,RCOOH \\ | & & & & | & & \text{fatty acid} \\ CH_2O.COR & & & & CH_2OH & & \\ \text{triacly glycerol} & & & & \text{glycerol} & & \end{array}$$

The important types of hydrolases are esterases (e.g. lipase), glycosidase, peptidases, phosphatases, thiolases, phospholipases, amidases, deaminases and ribonucleases.

4. Lyases. This group of enzymes catalyses the addition of groups to double bonds or catalyses cleavage by electronic rearrangement. For example, a decarboxylase catalyses the removal of CO_2 (a cleavage reaction) from keto acid;

$$R.CO.COOH \rightarrow RCHO + CO_2$$

5. Isomerases. These enzymes catalyse isomerization of several types. They include cis-trans, keto-enol and aldose-ketose interconversions. For example, and epimerase which catalyses the interconversions. For example, an epimerase which catalyses the conversion of D-xylulose 5-phosphate to D-ribulose phosphate, catalyses the transfer of group at asymmetric carbon atom;

Racemases, epimerases and some mutase are examples of isomerases.

$$\begin{array}{ccc} CH_2OH & & CH_2OH \\ | & & | \\ C=O & & C=O \\ | & & | \\ HO—C—H & & H—C—OH \\ | & \rightleftarrows & | \\ H—C—OH & & H—C—OH \\ | & & | \\ CH_2O\,(P) & & CH_2O\,(P) \\ \text{D-xylulose 5-phosphate} & & \text{D-ribulose 5-phosphate} \end{array}$$

6. Ligases. These enzymes catalyse bond formation and are accompanied by the hydrolysis of ATP or other nucleoside phosphate. These enzymes are also called synthetases. Some carboxylases are also kept in this category. For example, glutamine synthetase catalyses the formation of glutamine from glutamate and NH_3, involving ATP;

$$\begin{array}{lcl} \mathrm{COOH} & & \mathrm{COOH} \\ | & & | \\ \mathrm{CHMH_2} & & \mathrm{CHMH_2} \\ | & & | \\ \mathrm{CH_2} & + \mathrm{NH_3} + \mathrm{ATP} \rightleftarrows & \mathrm{CH_2} + \mathrm{ADP} + \mathrm{Pi} + \mathrm{H_2O} \\ | & & | \\ \mathrm{CH_2} & & \mathrm{CH_2} \\ | & & | \\ \mathrm{COOH} & & \mathrm{COOH_2} \\ \textbf{glutamate} & & \textbf{glutamine} \end{array}$$

ANTIENZYMES

The wall of the alimentary canal is not digested itself by its own proteolytic enzymes. Similarly, intestinal parasites such as round and hook worms, etc., are not digested by the proteolytic enzymes present in the alimentary canal. This has led to the discovery of what are known as **antienzymes**. It has been reported that such tissues, or worms posses certain proteins having antagonistic property to the proteolytic enzymes. Reports are also available regarding production of antienzymes in the serum when certain enzyme preparations are given parenterally. The mucous membrane is also believed to contain anti-proteolyic enzymes. *Ascaris* has been shown to contain **anti-trypsin, anti-proteolytic enzymes.** The well known antienzymes are **anti-pepsin, anti-trypsin, anti-rennin** and **anti-urease.**

ISOENZYMES (ISOZYMES)

Many enzymes within the same species or even within a single cell can exist in multiple molecular forms under normal condition and these are known as *isozymes*. This is in contrast to enzymes catalyzing the same reaction in different species which are called alloenzymes. Isozymes can differ in number of physical and chemical properties like net charge, resistance to denaturing agents, susceptibility to inhibitors and the kinetic constants *Km* and V*max*. However, they catalyze the same chemical reaction. Isozymes can arise due to genetically determined differences in primary structure as in the case of mitochondrial and cytosolic malate dehydrogenases. Enzymes derived from a single precursor, like α and Π -chymotrypsins are also isoenzymes. Monomer, oligomer and polymer forms of an enzyme can also be classified as isozymes. A well known example is liver glutamate dehydrogenase which exists as a hexamer usually, made up of identical polypeptide chains of Mr 60 kDa. The reaction catalyzed by this enzyme is shown below.

$$\text{L – Glutamate} \xrightarrow{-2H} \alpha\text{-Oxoglutarate} + NH_4^+$$

The monomeric form alone can act on L-alanine. Another cause for existence of isozymes is due to the differences in carbohydrate chains in enzymes which are glycoproteins. It is believed that isozymes of alkaline phosphatase arise due to this property.

Isozymes also arise in oligomeric enzymes made up of more than one type of polypeptide chain. A classical example is with

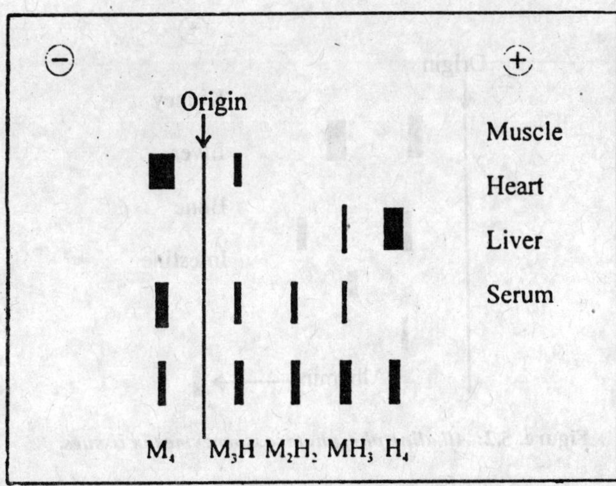

Figure. 5.1: ***Lactate dehydrogenase (LDH) isozyme patterns in tissues based on zone electrophoresis at pH 8.0.***

reference to locate dehydrogenase. Vertebrates possess two distinct genes producing two different polypeptide chains each with an Mr of 35 kDa. The M (muscle) and H (heart) froms can aggregate into active tetramers to yield five isozymes namely, M_4, M_3H, M_2H_2, MH_3 and H_4, In heart tissue, H_4 is the dominant form whereas in skeletal muscle, the M_4 form predominates. H_4 is powerfully inhibited by pyruvate the substrate whereas, M_4 form is only weakly inhibited. This is in accordance with the differential metabolism of pyruvate in the two tissues. In aerobic heart tissue, pyruvate takes an alternate route to form acetyl CoA whereas in anaerobic muscle it is converted to lactate. The M polypeptide is basic in nature whereas the H chain is acidic. This property can be used for the separation of lactate dehydrogenase isoenzymes by electrophoresis.

Each tissue has a characteristic pattern of isozymes. The relative proportions of M_4 (LDH–5), M_3H (LDH–4), M_2H_2 (LDH–3), MH_3 (LDH–2) and H_4 (LDH–1) in human serum in healthy persons are 5.0%, 8.0%, 27.0%, 35.0% and 25.0% respectively.

Isozymes of alkaline phosphatase and creatine kinase can also be separated by electrophoresis.

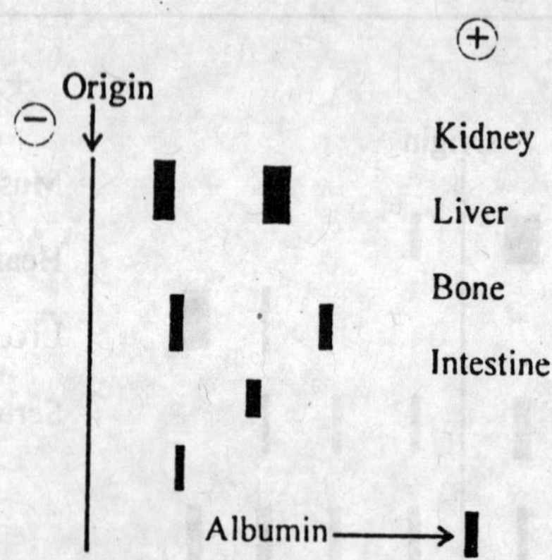

Figure. 5.2: ***Alkaline phosphatase isoenzymes in tissues.***

UDP – Galactose + N – Acetylglucosamine $\xrightarrow{\textit{Galactosyl transferase}}$ *Galactose – N–acetylglucosamine*

(UDP : Uridine diphospho)

In the presence of the protein α-lactalbumin, the modifier protein, glucose acts as the acceptor.

$$\text{UDP} - \text{Galactose} + \text{Glucose} \xrightarrow{\text{Calactosyl transferase: α-lactalbumin}} \text{Galactose} - \text{glucose.}$$

During pregnancy galactosyl transferase is synthesized. At the time of child-birth amount of α-lactalbumin is synthesized.

Biological significance

The study of isozymes provides useful and valuable information:

1. They are of great fundamental significance in the investigation of the molecular basis of cellular differentiation and morphogenesis.
2. They are proved to be excellent markers for gene activity and permit the identification of biochemical phenotypes.
3. They have made an important contribution to the dogma of genetics of one gene-one enzyme to one gene-one polypeptide.
4. They have many practical application:
   (a) In the diagnosis of heart attack-injured heart muscle releases characteristic patterns of isozymes into the blood stream and these may be readily detected on zymograms of blood. Both lactate dehydrogenase and creatine kinase isozymes are found useful in this diagnosis.
   (b) In the clinical diagnosis of diseases-the measurement of activities of enzymes such as lipase, phosphatases, lactate dehydrogenase, creatine kinase, trypsin, pepsin, etc. are found useful.

Some to the best studied isozymes include that of human lactate dehydrogenases (LDH), esterases, phosphatases, etc.

ISOLATION AND PURIFICATION OF ENZYMES

Since enzymes are very unstable molecules, their isolation from tissues is to be done under controlled conditions of pH, ionic strength and temperature. The standard procedures for extracting and purifying and enzyme are same as for any other protein. Enzyme assay is performed at each step of purification, so as to ensure that enzyme is not lost during purification. Following steps are usually followed in isolation and purification enzymes:

1. Extraction: The plant or animal tissue from which the enzyme is to be extracted is homogenised in a tissue homogeniser, a blender or with a mortar and pestle. Generally the extraction is done under cold (below 4°C) conditions because most of the enzymes are thermo-labile and they get inactivated at higher temperatures. The extraction is done with a buffer of specific ionic strength and pH. Ethylene diamine tetra acetic acid (EDTA) is generally added in the extraction medium to solubilize the membrane and to chelate heavy metals which could otherwise inhibit enzyme activity,. Detergents such as Triton-x are also included sometimes to solubilize the membranes. Many enzymic proteins contain disulfide (S—S) bonds due to the presence of sulfur containing amino acids, which are easily oxidized during enzymes extraction leading to the loss of enzyme structure and activity. To overcome this problem thiols such as mercaptoethanol or amino acid cysteine is added. Some times proteins such as casein or bovine serum albumin are also added in the extraction medium. These proteins protect the enzyme from the hydrolytic action of endogenously present proteases, which become active during tissue homogenisation.

2. Filtration and Centrifugation. The tissue extract is filtered using cheese cloth or filter paper to remove cell debris and fibers etc. The clear supernatant is then centrifuged in cold using a refrigerated centrifuge. The centrifugation can be used for purification as well as characterization of proteins. The techniques operates on the principle that molecules of different masses and shape move through a solution at different rates in the presence of the applied gravitational field, obtained by spinning the rotor of the centrifuge. High speed centrifugation (around 20000 × g) removes cell debris and also the bigger cell organelles such as chloroplasts and mitochondria. A very high speed centrifugation in ultra-centrifuge at about 100,000 × g removes all the cell organelles and only cytosolic proteins remain the supernatant. This supernatant can be used for further purification of cytosolic enzymes.

3. Precipitation. Like any other protein, enzymes are also highly charged molecules and can be precipitated out by proper charge breaking chemicals. Once their charges are neutralized, they form aggregates and settle down as precipitates. Acids, bases, salts and organic slovents are often used to precipitate out proteins. Protein or enzyme precipitation upon addition of concentrated solution of salt is called **salting out.** It is a complex process and involves disruption of various physical forces involved in protein solubilization. The added salt alters the structure of the solvent which can lead to large changes in protein configuration by altering the electrostatic interaction of charged groups on protein surfaces and the solvation of polar uncharged residues exposed to the solvent. The salt may also interfere with the formation of Vander Waal's forces between hydrophobic groups of the amino acids. Further, it may compete with the protein molecules for solvent molecules and thereby lower its solvation. In other word, salt dehydrates protein molecules which then form aggregates and precipitate out. For enzymes, the most commonly used salt precopotant is ammonium sulfate. The salt either as solid powder or as saturated solution is added slowly to the enzyme preparation to achieve desired concentration of the salt. Most proteins are precipitated out between 30 to 70% of ammonium sulfate. The enzymes settle down as precipitate and can be removed by centrifugation. The precipitate is dissolved in small amount of buffer of desired pH and ionic strength and is dialysed to remove excess of ammounium sulfate and other low molecular weight contaminants. About four folds enzyme purification is typically obtained with ammonim sulfate precipitation.

Some organic solvents such as acetone, methanol and ethanol are also used for enzyme precipitation. They are cooled up to – 40°C before their use and precipitation is carried out at O°C. They are added drop by drop to avoid local concentration.

4. Purification of Enzymes. The dialysed enzyme preparation is purified further by a variety of techniques. However, the most commonly used techniques are; chromatography and electrophoresis.

Adsorption or column chromatography is used to separate other proteins from the specific enzyme. Several spleen enzymes of human beings have been separated by this method. The column is packed with hydroxypetite and the enzyme is eluted with phosphate buffer of specific pH and ionic strength. Filtering the enzyme preparation through a column packed with sephadex or any other gel is also used to purify enzymes. This process is called molecular exclusion chromatography or molecular sieve chromatography.

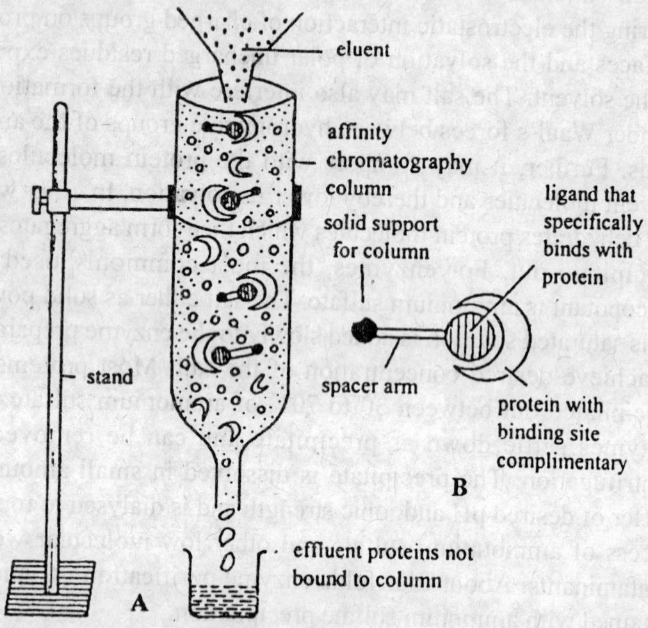

Figure. 5.3: ***Affinity chromatography. (The protein to be purified binds specifically to a ligand attached to the solid support in column. Other proteins run off. The protein bond in the column is then eluted with a suitable reagent).***

However, many enzymes are purified using affinity chromatography. In this procedure the enzymes are separated according to their biological specificity. The procedure is similar to that of column chromatography. The column is packed with inert material plus with a small molecular weight compound which acts as a ligand to bind with the enzymic protein. Generally the solid support and ligand are separated through a spacer arm (chemical). When the enzymes along with other proteins is poured through such a column the enzymic protein binds with the ligand (which is specific for the given enzyme only) and other proteins run off. The bound enzyme then can be recovered by using a suitable eluent, which will break the link between ligand and the protein. The protein (enzyme) will then elute and can be collected from the bottom of the column.

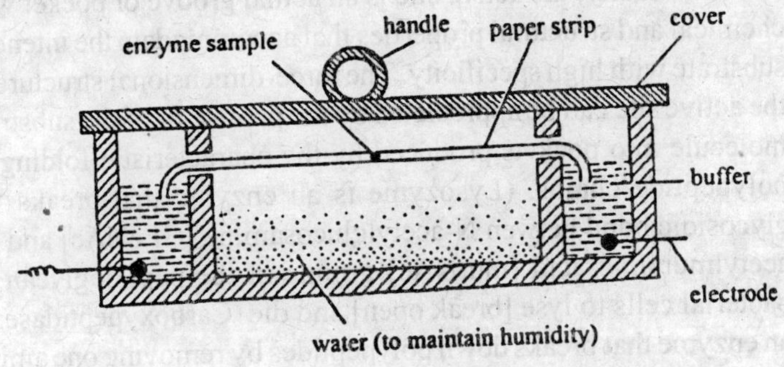

Figure. 5.4: ***Apparatus for paper electrophoresis***

Electrophoretic method is also often used for purifying enzymes and other proteins. It is also used for separating amino acids, nucleosides and nucleotides and also for various isozymes of the same enzyme. There are several types of electrophoretic methods, depending upon the material used for separating proteins. In all however, the principle of separation is base on charge on protein molecules. They are separated according to the charge on them, in a weak electric field. The simplest type of the electrophoresis is the paper electrophoresis in which a paper is used as a support for separating proteins. The protein sample is spotted on a paper strip. The ends of the paper strip are dipped in two electrode tanks filled a phosphate buffer of specific pH and ionic strength. The electrodes are connected to D.C. power supply of about 6 to 12 volts. When the current is applied to the electrophoresis apparatus through these electrodes, different proteins separate according to their charge (electrophoretic mobility) on the paper. They can be then located and identified using suitable colour reaction of the protein and biological assay.

THE ACTIVE SITE

One of the most important concepts to emerge from our understanding of enzymes as proteins is the active site. Every enzyme, regardless of the reaction it catalyzes or the details of its structure, contains somewhere within its tertiary configuration a characteristic cluster of amino acids forming the active site where the actual catalytic event, for which that enzme is responsible, occurs. Usually, the active site is an actual groove or pocket with chemical and structural properties that accommodate the intended substrate with high specificity. The three-dimensional structure of the active site can be appreciated by the precise fit of the substrate molecule into pockets produced by the characteristic folding of polypeptide chains. (Lysozyme is an enzyme that breaks the glycosidic bond betwen N-acetylglucosamine [GLcNAc] and N-acetylmuramic acid [murNAc] groups in the peptidoglycan of bacterial cells to lyse [break open] and die. Carboxypeptidase. is an enzyme that breaks down polypeptides by removing one amino acid at a time from the C-terminus of the polypeptide.

The amino acids that make up the active site of an enzyme are not usually contiguous to one another along the primary sequence of the protein. Instead, they are brought together in just the right conformation by the specific three-dimensional folding of the polypeptide chain. For the carboxypeptidase molecule, for example, the active site consists of a tightly bound zinc ion bonded to the side chains of the histidine at position 69, the glutamate at position 72, and another histidine at position 196 of the polypeptide chain. When the substrate for the enzymes binds to the active site, the structure of the enzyme changes. bringing the argining at position 145, the glutamate at position 270, and the tyrosine at position 248 into close proximity with the substrate. Each ot these amino acids plays a critical role in the actual catalytic process.

For carboxypeptidase, then 6 animo acids out of a total chain length of 307 are actually involved at the active site. This is typical of most enzymes, with the active site usually involving about 5% of the surface area of the enzyme. Also typical is the involvement of amino acids from distant positions along the primary structure of the polypeptide. This involvement underscores the importance of the overall tertiary structure of the protein; only as an enzyme molecule attains its stable three dimensional conformation are the specific amino acids brought together to constitute the active site.

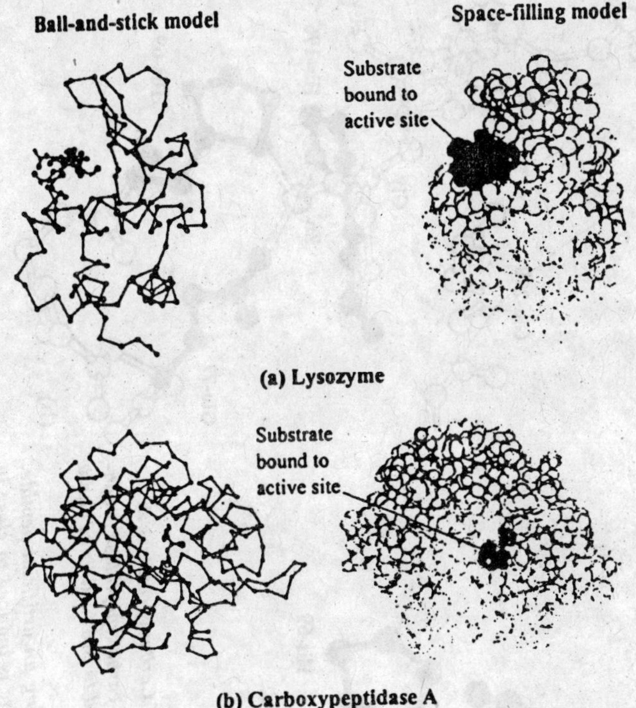

Figure. 5.5: *Molecular Structures of Lysozyme and Carboxypeptidase A.*

(a) *The enzyme lysozyme is represented as a ball-and-stick model (left), showing only the backbone of the protein as deducted by X-ray diffraction studies of protein crystals, and as a space-filling model (right) generated by computer. The molecules shown here is lysozyme from chicken eggs. A substrate molecule is shown in the active site, which appears as a cleft in the side of the enzyme molecule.*

(b) *The anzyme carobxypeptidase A is represented as a ball-and-stick model (left) and a space-filling model (right), with a substrate molecule bound to the active site.*

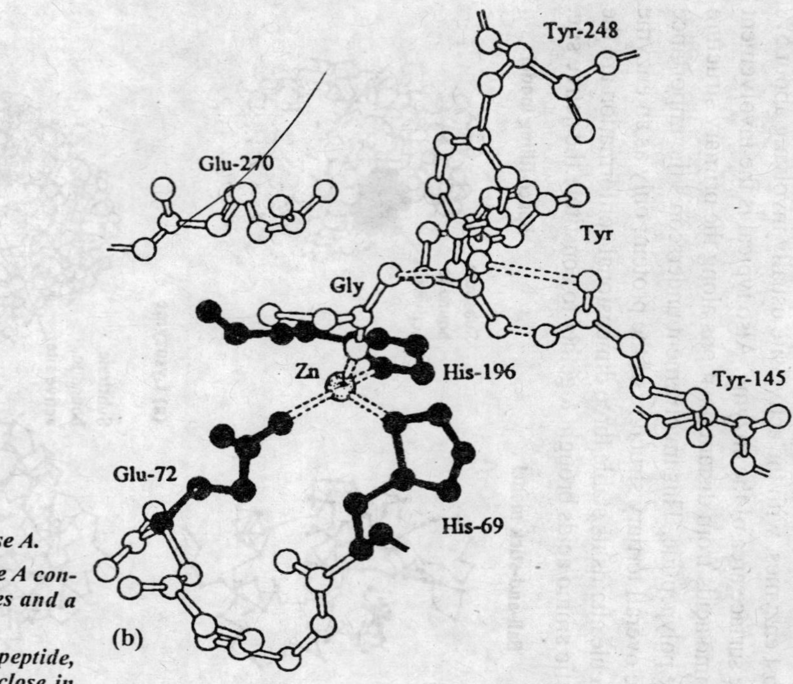

Figure. 5.6: *The Active Site of Carboxypeptidase A.*

(a) The unoccupied active site of carboxypeptidase A consists of a zinc ion coordinated to two histidines and a glutamate.

(b) Upon substrate binding (using an artificial peptide, shown in pink), the active site is induced to close in on the substrate, bringing an arginine, a glutamate, and a tyrosine into close proximity with the substrate.

Of the 20 different amino acids that make up proteins, only a few are actually involved in the active sites of the many proteins that have been studied.In most cases, these are the amino acids cysteine, histidine, serine, aspartate, glutamate, and lysine. All of these can participate in binding or bonding of the substrate to the active site during the catalytic process, and several (histidine, aspartate, and glutamate) also serve as donors or acceptors of protons.

Some enzymes consist not only of one or more polypeptide chains but of specific nonprotein components as well. These components are called **prosthetic groups** and are usually either small organic molecules or metal ions, such as the iron in catalase. Frequently, they function as electron acceptors. Where present, prosthetic groups are located at the active site and are indispensable for the catalytic activity of the enzyme. Carboxypeptidase A is an example of an enzyme with a prosthetic group, containing a zinc atom at its active site.

ENZYME SPECIFICITY

A consequence of the structure of the active site is that enzymes display a high degree of substrate specificity, evidenced by an ability to discriminate between very similar molecules. Specificity is probably one of the most characteristic properties of living systems, and enzymes are especially dramatic examples of biological specificity.

We can illustrate their specificity by comparing enzymes with inorganic catalysts. Most inorganic catalysts are quite nonspecific in that they will act on a variety of compounds that share some general chemical feature. Consider, for example the *hydrogenation* of (addition of hydrogen to) an unsaturated C = C bond:

$$R{-}\overset{\displaystyle H}{\overset{|}{C}}{=}\overset{\displaystyle H}{\overset{|}{C}}{-}R' + H_2 \qquad R{-}\underset{\displaystyle H}{\underset{|}{\overset{\displaystyle H}{\overset{|}{C}}}}{-}\underset{\displaystyle H}{\underset{|}{\overset{\displaystyle H}{\overset{|}{C}}}}{-}R' + H_2$$

This reaction can by carried out in the laboratory using a platinum (Pt) or nickel (Ni) catalyst, as indicated, These inorganic catalysts are very nonspecific, however; they can by used to hydrogenate a wide variety of unsaturated compounds. In fact, nickel or platinum is used commercially to hydrogenate polyunsaturated vegetable oils in the manufacture of solid cooking fats or shortenings. Regardless of the exact structure of the unsaturated compounds, it can be effectively hydrogenated in the presence of nickel or platinum.

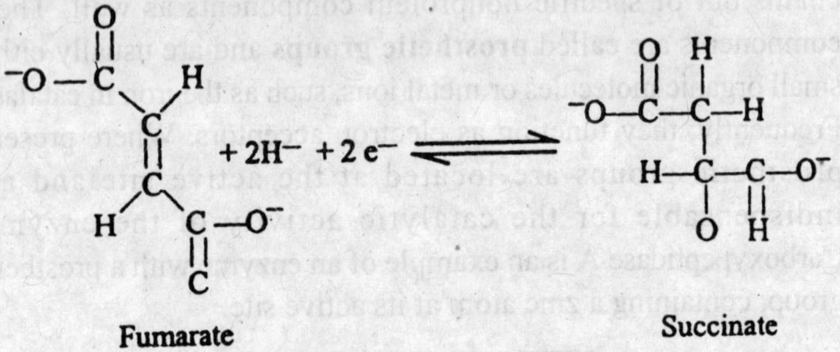

$$\text{Fumarate} + 2H^- + 2e^- \rightleftharpoons \text{Succinate}$$

Fumarate **Succinate**

This particular reaction is catalyzed in cells by the enzyme *succinate dehydrogenase* (so named because it normally functions in the opposite direction during energy metabolism). This dehydrogenase, like most enzymes, is highly specific. It will not add or subtract hydrogens from any compounds except those shown in reaction 6–4. In fact, this particular enzyme is so specific that it will not even recognize maleate, a geometric stereoisomer of fumarate.

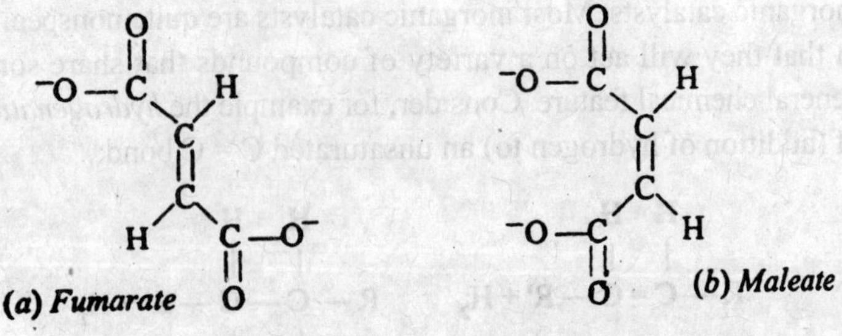

(a) *Fumarate* **(b) *Maleate***

Figure. 5.7: ***The Stereoisomers (a) Fumarate and (b) Maleate***

Not all enzymes are quite this specific; some accept a number of closely related substrates, and other accept any of a whole group of substrates as long as they possess some common structural feature. Such **group specificity** is seen most often with enzymes involved in the synthesis or degradation of polymers. The purpose of carboxypeptidase A is to degrade polypeptide chains from the carboxylend; thus, it makes sense for the enzyme to accept any of a wide variety of polypeptides as substrate, since it would be needlessly extravagant of the cell to require a separtate enzyme for every different peptide bond that has to be hydrolyzed in polypeptide degradation.

In general, however, enzymes are highly specific with respect to substrate, such that a cell must possess almost as many different kinds of enzymes as it has reactions to catalyze. For a typical cell, this means that several thousand different kinds of enzymes are necessary to carry out its full metabolic program. At first, that may seem wasteful in terms of proteins to be synthesized, genetic information to be stored and read out, and enzyme molecules to have on hand in the cell. But you should also be able to see the tremendous regulatory possibilities this suggests, a point we will return to later.

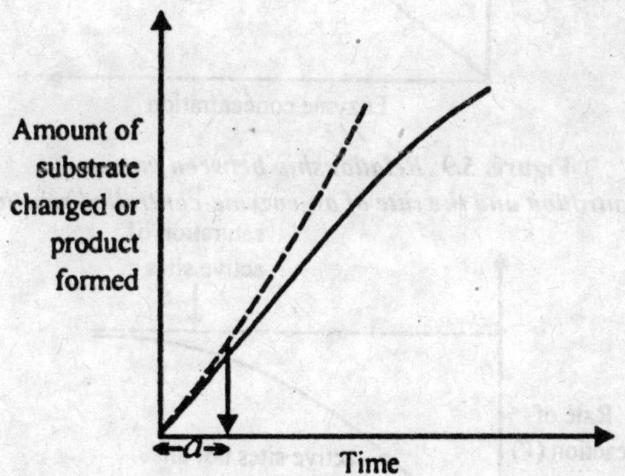

Figure. 5.8: ***The rate of an enzyme-controlled reaction***

Enzyme concentration

Provided that the substrate concentration is maintained at a high level, and other conditions such as pH and temperature are kept constant, the rate of reaction is proportional to the enzyme concentration. Normally reactions are catalysed by enzyme concentrations which are much lower than substrate concentrations. Thus as the enzyme concentration is increased, so will be rate of the enzyme reation.

SUBSTRATE CONCENTRATION

For a given enzyme concentration, the rate of an enzyme reaction increases with increasing substrate concentration. The theoretical maximum rate (V_{max}) is never quite obtained, but there comes a point when any further increase in substrate concentration produces no significant change in reaction rate. This is because at high substrate concentration the active sites of the enzyme molecules at any given moment are virtually saturated with substrate. Thus any extra substrate has to wait until the enzyme/ substrate complex has released the products before it may itself enter the active site of the enzyme.

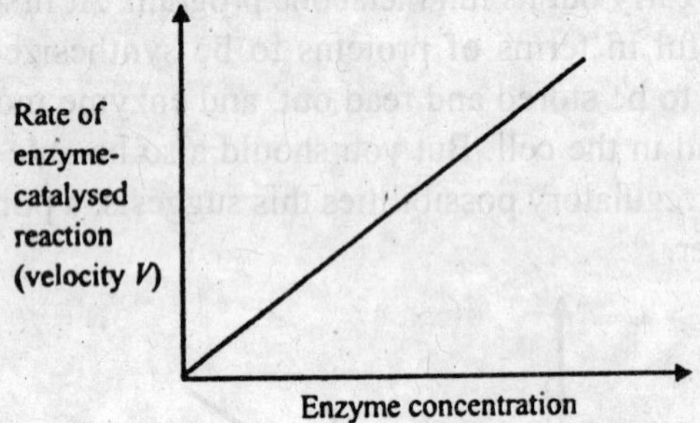

Figure. 5.9: ***Relationship between enzme concentration and the rate of an enzyme-controlled reaction.***

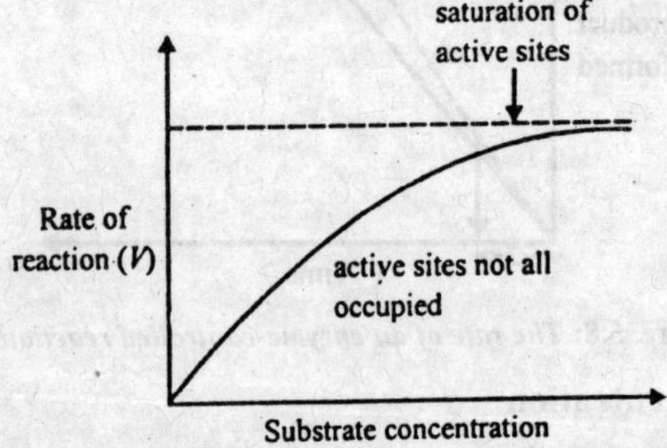

Figure. 5.10: ***Effect of substrate concentration on the rate of an enzyme-controlled reaction.***

TEMPERATURE

Heating increase molecular motion. Thus the molecules of the substrate and enzyme move quickly and chances of their bumping into each other are increased. As a result there is a greater probability of a reaction occuring. The temperature that promotes maximum activity is referred to as the optimum temperature. If the temperature is increased above this level, then a decrease in the rate of the reaction occurs despite the increasing frequency of collisions. This is because the secondary and tertiary structures of the enzyme have been disrupted, and the precise structure of the active site is gradually lost. The bonds which are most sensitive to temperature change are hydrogen bonds and hydrophobic interactions.

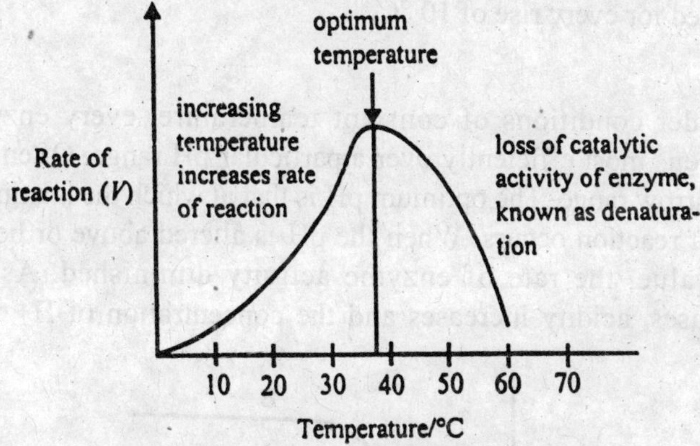

Figure. 5.11: ***Effect of temperature on the rate of an enzyme-controlled reaction.***

Most mammalian enzymes have a temperature optimum of about 37–40°C but enzymes with higher optima exist. For example, the enzymes of bacteria living in hot springs may have an optimum temperature of 70°C or higher. Such enzymes have been used in biological washing powders for high temperature washed. If temperature is reduced to near or below freezing point, enzymes are **inactivated,** not denatured. They will regain catalytic influence when higher temperatures are restored.

Today techinques of quick – freezing food are in widespread

use as a means of preserving food for long periods. This not only prevents growth and multiplication of microrganisms, but also deactivates their digestive enzymes thus making it impossible for them to decompose food. The natural enzymes in the food itself are also inactivated. However, once frozen, it is necessary to keep the food at subzero temperatures until it is to be prepared for consumption.

TEMPERATURE COFFICIENT, Q_{10}

The effect of temperature on the rate of a reaction can be expressed as the temperature coefficient, Q_{10}.

Q_{10} = rate of reaction at (x + 10) °C/rate of reaction at x°C

Over a range of 0–40°C, Q_{10} for an enzyme–controlled reaction is 2. In other words, the rate of an enzyme–controlled reaction is doubled for every rise of 10 °C.

pH

Under conditions of constant temperature, every enzyme functions most efficiently over a particular pH range. Often this is a narrow range. The optimum pH is that at which the maximum rate of reaction occurs. When the pH is altered above or below this value, the rate of enzyme activity diminished. As pH decreases, acidity increases and the concentration of H+ ions

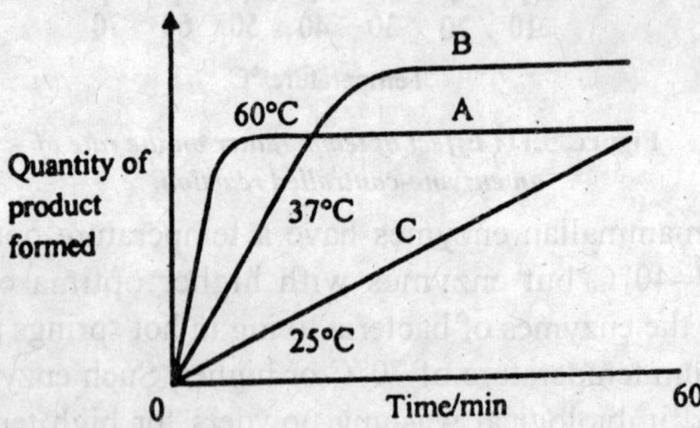

Figure. 5.12: ***Time course of an enzyme reaction at various temperatures.***

increases. This increases the number of positive charges in the medium. Change in pH alter the ionic bonding that helps to maintain the specific shape of the enzyme. Thus the pH change laeds to an alteration of enzyme shape, including its active site. If extremes of pH are encountered by an enzyme, then it will be denatured.

TABLE 5.1: OPTIMUM PH VALUES FOR SOME ENZYMES.

Enzyme	Optimum pH	Enzyme	Optimum pH
Pepsin	2.00	Catalase	7.60
Sucrase	4.50	Chymotrypsin	7.00–8.00
Enterokinase	5.50	Pancreatic lipase	9.00
Salivary amylase	4.80	Arginase	9.70

Biomolecules

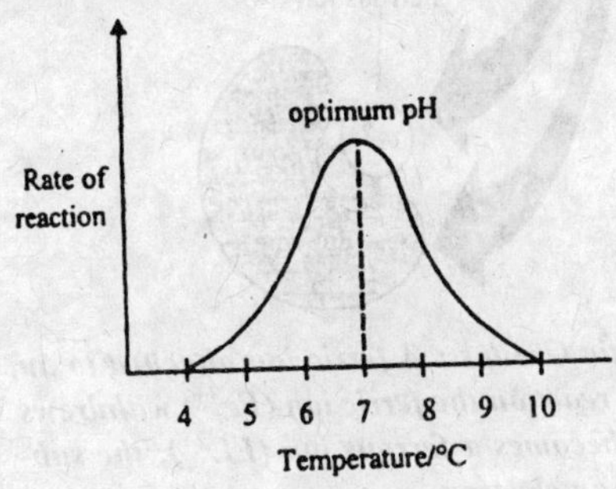

Figure. 5.13: ***Effect of pH on the rate of an enzyme controlled reaction.***

Regulation of Enzyme Activity

Various substances, some occurring naturally in cells and others artificial, act upon enzymes to increase or decrease the rates of enzyme catalyzed reactions. Those that occur naturally regulate metabolism; the artificial ones are used either to treat disease or to study how enzymes work. Some substances, called **inhibitors,** that inhibit enzyme–catalyzed reactions produce reversible effects—that is, these inhibitors can become unbound. Enzymes consisting of multiple subunits are subject to another type of control called allosteric regulation.

IRREVERSIBLE INHIBITION

Some inhibitors irreversibly modify certain side chains at active sites, ruining the enzymes by destroying their capacity to function as catalysts. An example of such an **irreversible inhibitor** is DFP

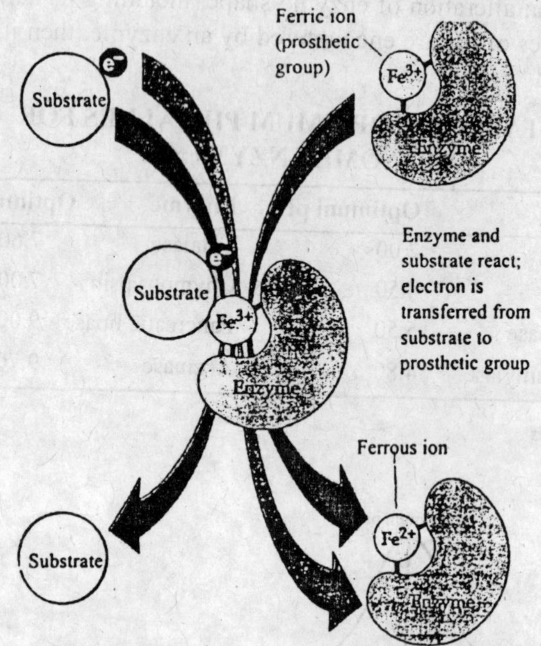

Figure. 5.14: ***Metal Ions as Prosthetic Groups : A ferric ion attached to an enzyme as a prosthetic group. In this reaction the ferric ion (Fe^{3+}) withdraws the electron from the substrate and becomes a ferrous ion (FE^{2+}); the substrate moves on, altered by the loss of an electron.***

(disopropylphos-phorofluoridate). DFP reacts with a hydroxyl group belonging to the amino acid serine at the active site of the enzyme trypsin, preventing the use of this side chain in the catalytic machanism.

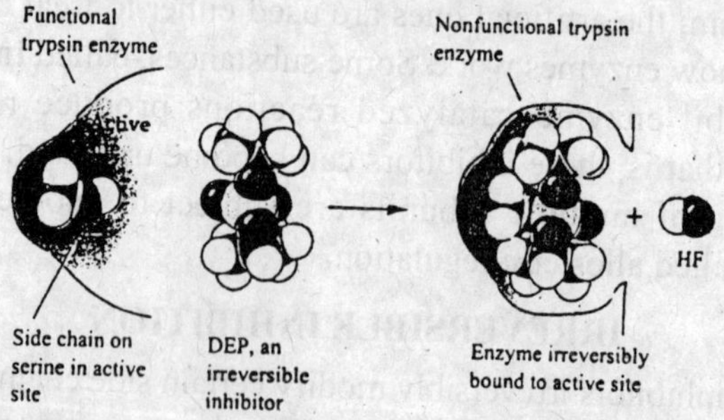

Figure. 5.15: ***Irreversible Inhibition DFP disables the digestive enzyme trypsin by reacting with a hydroxyl group belonging to the amino acid serine.***

DFP is an irreversible inhibitor not only for the protein-digesting enzyme trypsin but also for many other enzymes whose active sites contain serine. Among these is acetylcholinesterase, an enzyme that is essential for the orderly propagation of signals from one nerve cell to another. Because of their effect on acetylcholinesterase, DFP and other similar compounds are classified as nerve gases.

REVERSIBLE INHIBITION

Not all inhibitory action is irreversible. Some inhibitor molecules are similar enough to a particular enzyme's natural substrate to bind to the active site, yet different enough that the enzyme catalyzes no chemical reaction. When such a molecule is bound to the enzyme, the natural substrate cannot enter the active site; thus intruder effectively wastes the enzyme's time, inhibiting its catalytic action. These are called **competitive inhibitors** because they compete with the natural substrate for the active site and block it. The blockage is reversible, however; a competitive inhibitor may become unbound, leaving the active site unchanged. If enough natural substrate molecules are present, they can complete successfully with inhibitor molecules for empty active sites.

Consider the enzyme succinate dehydrogenase, which is subject to competitive inhibition. Recall that this enzyme, found in all mitochondria, removes two hydrogen atoms from succinate to produce fumarate, it then transfers the hydrogens to another molecules. The other molecules shown in the figure are competitve inhibitors of succinate dehydrogenase. They resemble succinae enough that the enzyme is fooled into binding them. Having bound them, however, the enzyme can do nothing more with them, because the inhibitors are the wrong size and shape or have key chemical group in the wrong places. The enzyme molecule cannot bind a succinate molecule until the inhibitor molecule has moved out of the active site.

Dissociation of the inhibitor does occur because binding of a competitive inhibitor is reversible, *as is binding of the substrate.* For example, when the competitive inhibitor malonate is added to a solution containing succinate and succinate dehydrogenase, the reaction of succinate to fumarate is slowed. The effect of malonate can be overcome, however, if enough succinate is added. The relative concentrations of substrate and inhibitor determine which of these is more likely to bind to the active site.

Inhibitors that do not react the active site are called **noncompetitive inhibitors.** Noncompetitive inhibitors bind to the enzyme at a site away from the active site. Their binding causes a conformational change in the protein that alters the active site. The active site still binds substrate molecules, but the rate of product formation is reduced. Noncompetitive inhibitors can become unbound, so their effects are reversible. Because they do not bind to the active side, their effects do not change as substrate concentration changes.

(a) Competitive inhibition

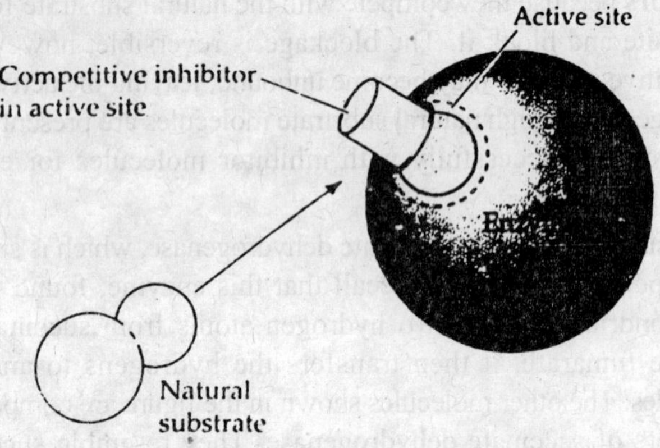

The enzyme molecule's function is disabled as long as the inhibitor remains bound but if the inhibitor becomes unbound, a substrate molecule may bind to the active site.

(b) Competitive inhibition of succinate dehydrogenase

The given series of molecules of increasing length compete with succinate for the enzyme's active site. The similarity that fits them all to the same active site is the presence of two negatively charged carboxyl groups, one at each end of the molecule.

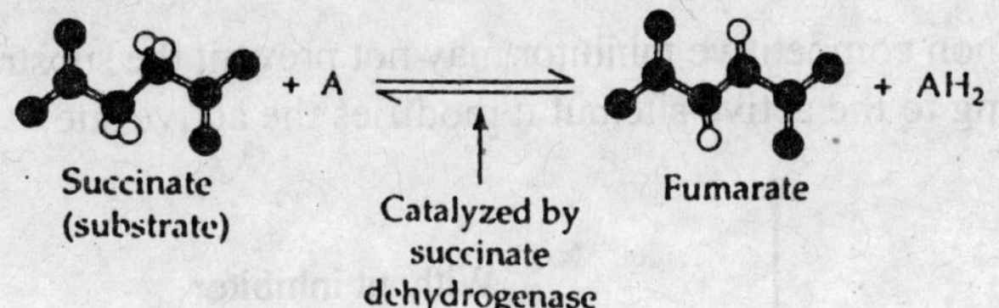
+ A
+ AH_2
Succinate
(substrate)
Catalyzed by
succinate
dehydrogenase
Fumarate

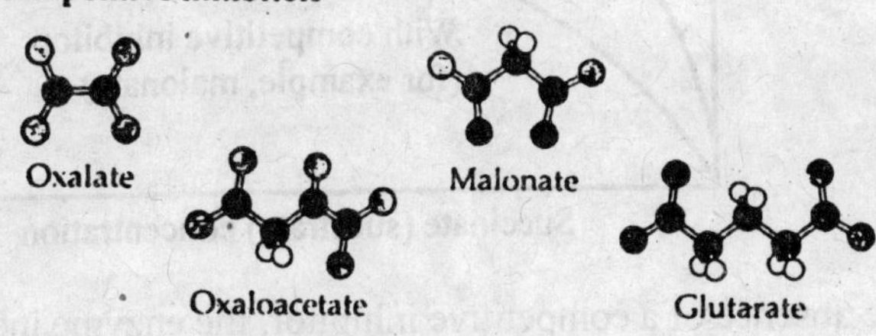
Competitive inhibitors
Oxalate
Malonate
Oxaloacetate
Glutarate

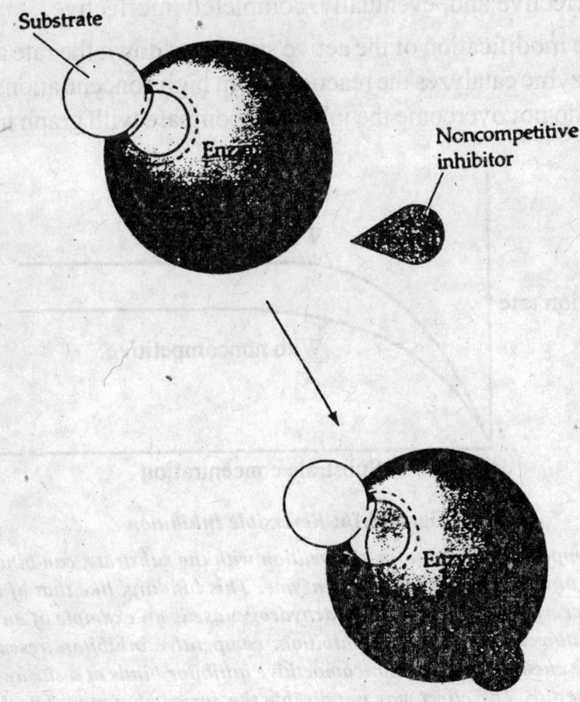
(c) Noncompetitive inhibition
Substrate
Noncompetitive
inhibitor

A non competitive inhibitor may not prevent the substrate from binding to the active site,but it modifies the active site.

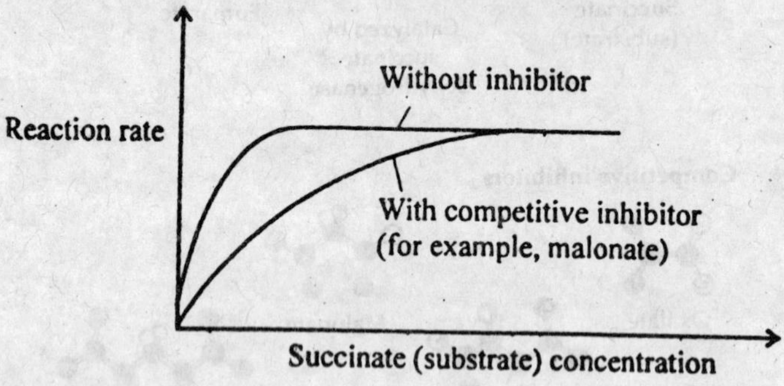

In the absence of a competitive inhibitor, the enzyme increases reaction rate more than it does when one is present. As substrate concentrations increase, however, competitive inhibition become less effective and, eventually, completely ineffective.

The modification of the active site slows down the rate at which the enzyme catalyzes the reaction. Even high concentrations of substrate do not overcome the inhibitor; compare with graph in (b).

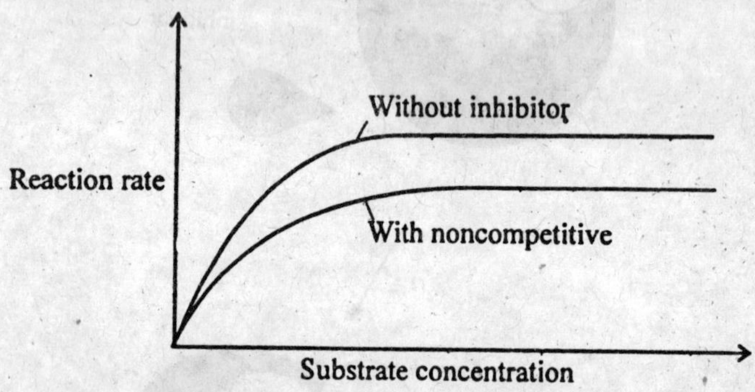

Figure 5.16: *Reversible Inhibition*

(a) A competitive inhibitor, in competition with the substrate, can bind the active site and thus can disable the enzyme. This binding, like that of the substrate, is reversible, (b) Succinate dehydrogenase is an example of an enzyme that is subject to competitive inhibition. competitive inhibitors resemble the substrate chemically. (c) A noncompetitive inhibitor binds at a site away from the active site. The effect may not disable the enzyme but may slow down the reaction.

ALLOSTERIC ENZYMES

Many important enzymes are larger and more complex than the ones we have discussed so far, which are individual polypeptides. These complex enzymes have quaternary structures consisting of two or more polypeptide subunits, each with a molecular weight in the tens of thousands.

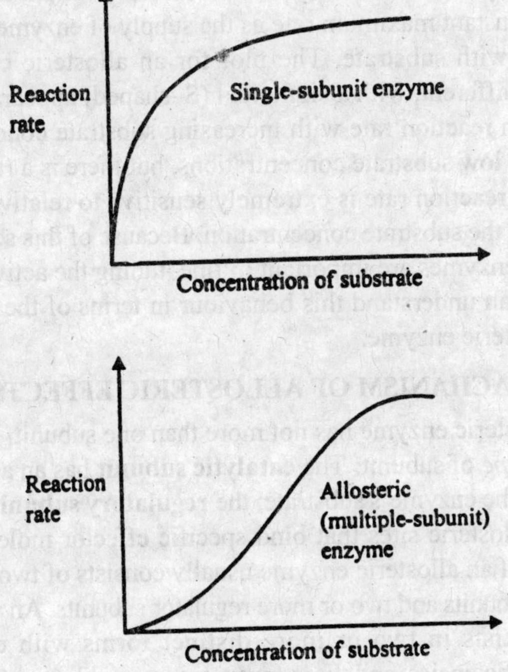

Figure. 5.17: ***Allostery and Reaction Rate.***

The activity of these complex enzymes is controlled by molecules, called **effectors,** that may have no similarity either to the reactants or to the products of the reaction being catalyzed. Effectors operate by binding to a site on the enzyme other than the active site. Binding at the **allosteric site** enhances or diminishes reactions at the active site; effectors thus can be activators or inhibitors of enzymes. Because many effector–substrate pairs differ structurally, this phenomenon is called allostery, meaning "different shape". Enzymes subject to allosteric control are called **allosteric enzymes;** all have two or more sub–units.

Allosteric enzymes and single–subunit enzymes differ greatly in their effects on reaction rates when the substrate concentration is low. Graphs of reaction rates plotted against substrate concentration show this difference. The reaction rate first increases very sharply with increasing substrate concentration, then tapers off to a constant maximum rate as the supply of enzyme becomes saturated with substrate. The plot for an allosteric enzyme is radically different, with a sigmoidal (S–shaped) appearance. The increase in reaction rate with increasing substrate concentration is slight at low substrate concentrations, but there is a range over which the reaction rate is extremely sensitive to relatively small changes in the substrate concentration. Because of this sensitivity, allosteric enzymes are important in fine–tuning the activities of a cell. We can understand this behaviour in terms of the structure of an allosteric enzyme.

MACHANISM OF ALLOSTERIC EFFECTS

An allosteric enzyme has not more than one subunit, but more than one *type* of subunit: The **catalytic subunit** has an active site that binds the enzyme's substrate; the **regulatory subunit** has one or more allosteric sites that bind specific effector molecules. A molecule of an allosteric enzyme usually consists of two or more catalytic subunits and two or more regulator subunits. An allosteric enzyme exists in two or more distinct forms with different catalyitc efficiencies, and these forms are in equilibrium with each other. In the simple cases we will examine, the **active form** has full catalytic activity, whereas the **inactive form** is totally without activity. In the active form, the active sites on the catalytic subunits can bind substrate and convert it to product. In the inactive form, the active sites have been distorted in such a way that cannot bind substrate; however, the allosteric sites are able to bind an effector, which we will consider in this example to be an inhibitor. The regulatory subunits of the active form of the enzyme have deformed allosteric sites and cannot bind effector. When neither substrate nor inhibitor is present, the active and inactive forms convert rapidly back and forth in equilibrium, the equilibrium constant being characteristic of the given enzyme.

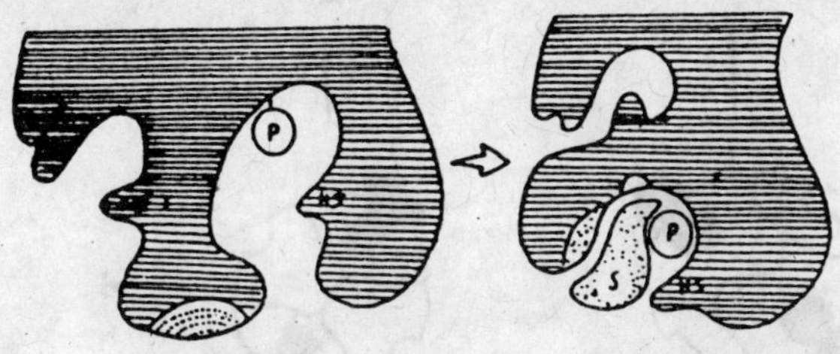

Figure. 5.18: ***Conformational change in the enzyme as a result of the entry of substrate (s) into the active site; according to Koshland.***

What happens when inhibitor or substrate is present? If substrate is present, some of the substrate binds to the active sites of active enzyme molecules; while the enzyme-substrate complex exists, those enzyme molecules cannot be converted to the inactive form. Inactive molecules, however, are being converted to active one at the same rate as before, so an increase in the concentration of active enzyme results from the presence of subtrate! In addition because each active enzyme molecule has two active sites, one enzyme molecule can bind two substrate molecules and simultaneously catalyze reactions of both of them. This explains the upward curvature at the lower left of a plot of reaction rate versus substrate concentration for an allosteric enzsyme. Increasing the substrate concentration increases the availabilty of active enzyme and of active and hence rapidly accelerates the reaction rate.

If inhibitor is present, the concentration of active enzyme *decreases* and the reaction is thus inhibited. The inhibitor binds to the allosteric site of the *inactive* form of the enzyme, preventing the conversion of the inactive to the active form; but the conversion of the active to the inactive enzyme is not affected by the inhibitor. The overall effect is to decrease the concentration of the active form and thus inhibit the enzymatic reaction. An allosteric activator works in a similar way, except that it binds to the regulatory subunit of the *active* form and holds it in the active configuration. Note that allosteric inhibitors and activators do not

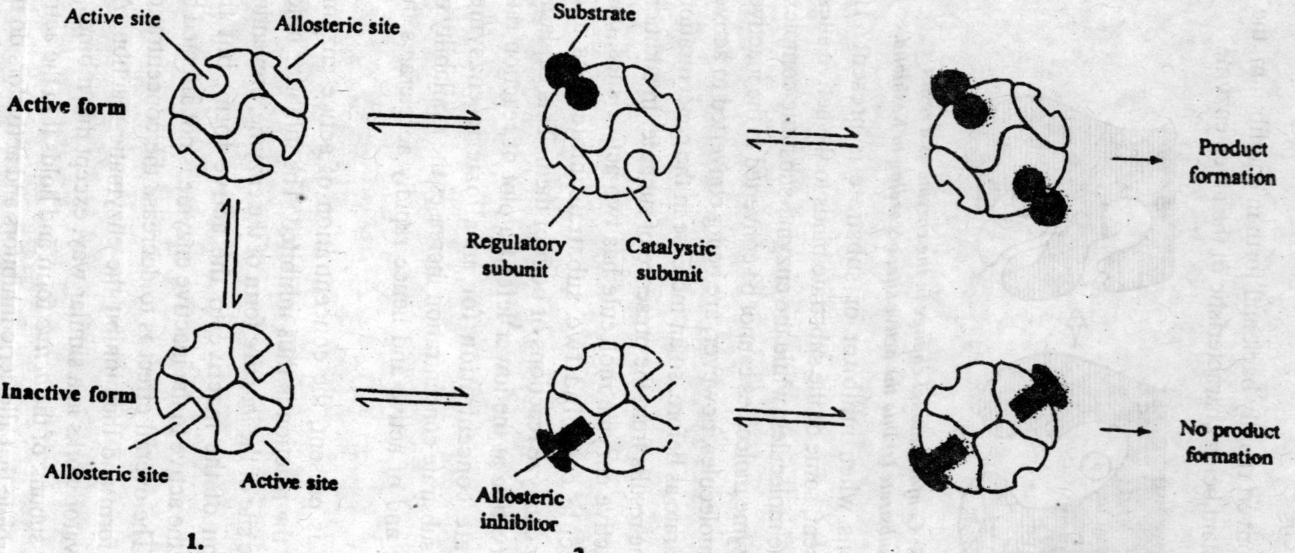

Figure. 5.19: *Allosteric Regulation of Enzymes*

The hypothetical enzyme shown her has four subunits, two catalutic (blue) and the other two regulatory. When the enzyme is in its active form the active sites on the catalytic subunits can accept substrate. When the enzynme is in its inactive form, the allosteric sites on the regulatory subunits can accept inhibitor.

modify the structure of the enzyme; rather, they interfere with its conversion to another form with which it is normally in equilibrium. This mechanism of allosteric inhibition and activation was proposed by the French molecular biologist Jacques Monod and his colleagues in 1956.

Control of Metabolism through Allosteric Effects

An organism's metabolism is the sum total of the biochemical reactions that take place with in it. These reactions proceed along **metabolic pathways,** which are sequences of enzyme-catalyzed reaction. In the sequences, the product of one reaction is the substrate for the next. Some pathways synthesize, step-by-step, the important chemical building blocks from which macromolecules are built; others trap energy from the environment; still others have different functions. Some metabolic pathways are branched. In a branched pathway, one or more of the intermediate substances are acted upon by more than one enzyme and thus are sent through more than one metabolic branch.

At the branching points where metabolic pathways diverge, **ragulatory enzymes** catalyze reactions. These regulatory enzymes are like switches. What flips such a switch? The end product of a branch pathway may block the initial step on that branch pathway, reducing the formation of the end product. This illustrates the principle of **negative feedback,** also evident, for example, in thermostats on furnaces.

The end product of a particular pathway typically is an allosteric inhibitor of the regulatory enzyme catalyzing the first **committed step** in its own synthesis—that is, the earliest step in the branched path way that leads only to the synthesis of that end product and no other. The first committed steps in metabolic pathways are particularly effective points for feedback control. For instance, inhibition of the B-to B step in Figure shunts all the reactants into the other branch of the pathway, whereas inhibition of the C-to-D reaction, one step later, leads only to a possibly harmful and certainly wasteful buildup of substance C. Living things geerally do not accumulate unneeded intermediates.

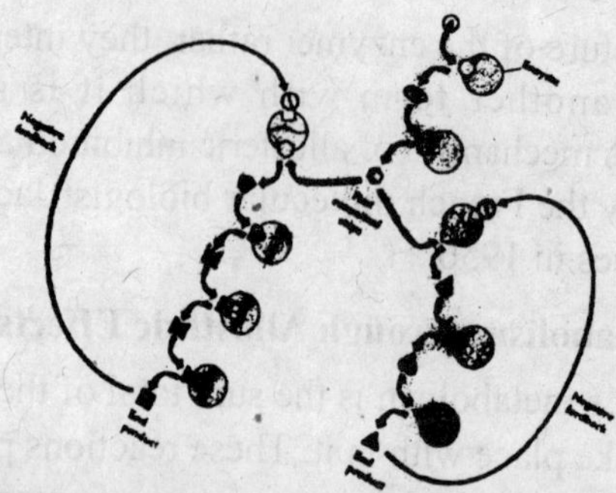

Figure. 5.20: ***Feedback in Metabolic Pathways.***
The reactions C to D and C to H are the first committed steps in the branch pathways leading to end products G and J, respectively. These end products can block the first committed steps by acting as allosteric inhibitors of the enzymes catalyzing the reactions. Thus, if levels of products G and J build up beyond what the cell needs for other reactions, the extra molecules provide the negative feedback that turns off the synthesis of G and J.

When two different end products, produced by different branches of a pathway, are both present in excesive concentration, they may act together to inhibit and earlier branch-point enzyme—one that catalyzes the first committed step for formation of these two products. **Concerted feedback inhibition** like this results in

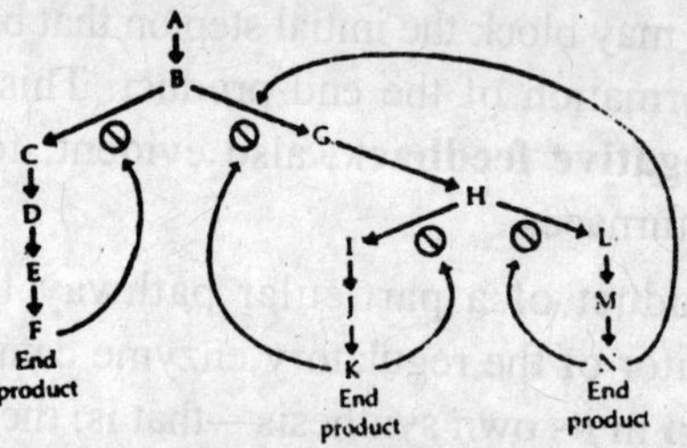

Figure. 5.21: ***Concerted Feedback Inhibition***
A variety of feedback controls come into play in branching metabolic path way. Inhibiting the B → C reactin, for example, shunts reactant B into the B → G step. In some cases in which a single compound (H) gives rise to more than one end product (K and N), both end products can join in concerted inhibition of the enzyme catalyzing the first reaction committed to the production of H (B → G); each end product may also inhibit its own branch pathway.

further efficiency: In this example intermediates G and H do not build up as they would if only the steps to I and L were inhibited by their individual end products. Concerted feedback inhibition requires that the enzyme have two allosteric sites, both of which must be bound to inhibitors to stop the enzyme's activity.

Allosteric regulation is very effective: It allows rapid adjustment to short-term changes in metabolism or in the environment. The activities of enzyme molecules are adjusted by their interactions with small molecules, the end products. If a particular enzyme is not needed, might it not be a good idea simply to stop making it until it is needed? Wouldn't it be advantageous to be able to regulate *production* as well enzyme *activity?* This is indeed the case, and the regulation of enzyme synthesis plays an important role in controlling development and metabolism.

CHEMICAL KINETICS

Before we proceed to examine catalysis of reactions by enzymes, some relationships and terms employed in measuring and expressing the rates of chemical reactions must be outlined. Chemical reactions may be classified on the basis of the number of molecules that must ultimately react to form the reaction products. Thus, we may have *monomolecular, bimolecular,* and *terimolecular* reactions.

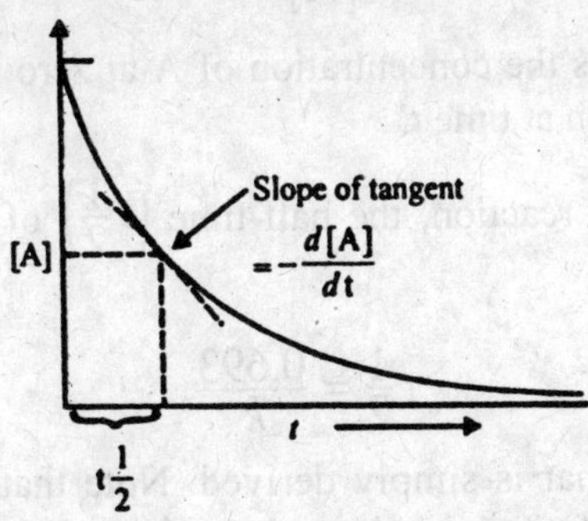

Figure. 5.22: ***Plot of the course of a first reaction. The half-time*** $t\left(\frac{1}{2}\right)$ ***is the time required for one-half of the initial reactant to be consumed.***

Chemical reactions may be classified on a kinetic basis, by *reaction order.* thus, we may have zero-order, first-order, second-order, and third -order reactions, depending on how the reaction rate is influenced by the concentration of the reactants under a given set of conditions.

First-order reactions are those which proceed at a rate exactly proportional to the concentration of *one* reactant. The simplest example is when the rate of the reaction

$$A \rightarrow P$$

is exactly proportional to the rate of disappearance of A (or the rate of appearance of P). For such a case, the reaction rate at any time *t* is given by

$$\frac{-d[A]}{dt} = k[A]$$

in which [A] is the molar concentration of A and $-d[A]/dt$ is the rate at which the concentration of A decreases. The proportionality constant k is called the *rate constant* or *specific reaction rate*. First-order rate constant have the dimensions of reciprocal time,i.e., sec^{-1} or min^{-1}.

The integrated form of this equation, which is more useful for carrying out kinetic calculation, is

$$\log_{t0} \frac{[A]}{[A]} = \frac{kt}{2.303}$$

in which $[A_o]$ is the concentration of A at zero time and [A] is the concentration at time *t*.

In first-order reaction, the half-time $\left(t\frac{1}{2}\right)$ of the reaction is given by

$$t\frac{1}{2} = \frac{0.693}{k}$$

a relationship that is simply derived. Note that in first–order reactions the half–time is independent of the initial concentration of substrate.

Second–order reaction are those in which the rate is proportional to the product of the concentrations of *two* reactants or to the second power of a single reactant. The simplest example is the reaction

$$A + B \rightarrow P$$

When the rate of this reaction, which may be designated as–(d[A]/dt), –(d [B] /dt), or + (d [P] / dt), is proportional to the product of the concentrations of A and B, we have

$$\frac{-d[A]}{dt} = k\,[A][B]$$

the second–order rate equation, in which k is the second–order rate constant. If the reaction has the from

$$2A \rightarrow P$$

and the rate is proportional to the product of the concentration of the two reacting molecules, the second-order rate equations is

$$\frac{-d[A]}{dt} = k\,[A][A] = k\,[A]^2$$

The rate constants of second–order reactions have the dimensions 1/concentration × time, or M^{-1}. The integrated form of the second-order expression is

$$t = \frac{2.303}{k([A_0]-[B_0])} \log_{t0} \frac{[B_0][A]}{[A_0][B]}$$

where$[A_o]$ and $[B_o]$ are initial concentrations and [A] and [B] the concentrations at time t.

It is important to note that the reaction

$$A + B \rightarrow P$$

which we have taken as an example, is not mecessarily a second–order reaction under all circumstances. Under some conditions this biomlecular reaction can be a first–order reaction. For example, if the concentration of A were very high and that of B very low, this reaction might be first–order with respect to B, since its rate is then proportional to the concentration of only one reactant (in this case, B).

This order of a reaction is thus determined by the conditions under which it is taking place and is not automatically a reflection of whether the reaction is monomolecular, bimolecular, or trimolecular.

Third–order reactions, which are relatively rare, are those whose velocity is proportional to the product of three concentration terms. Some chemical reactions are independent of the concentration of any reactant ; these are called *zero–order reactions.* Many catalyzed reactions are zero–order with respect to the reactants. When this is the case, the rate of reaction depends on the concentration of catalyst or on some factor other than the concentration of the molecular species undergoing reaction. Reaction rates need not necessarily be pure first–order; often mixed–order reactions are observed under certain conditions.

CATALYSIS

A chemical reaction such as A $\rightarrow$ P takes place because a certain fraction of the population of A molecules at any given

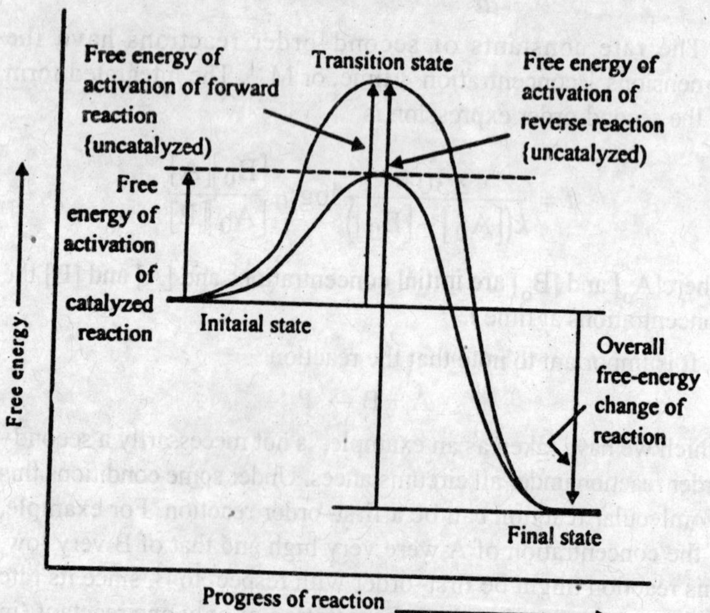

Figure. 5.23: ***Energy diagram for a chemical reaction, uncatalyzed and catalyzed.***

instant possesses much more energy than the rest of the population—enough energy to attain an "activated state," in which a chemical bond may be made or broken to form the product(s) P. The term *activation energy* refers to the amount of energy, in calories required to bring all the moleculs of 1 mole of a substance at a given temperature to this activated state.

In every chemical reaction there is a *transition state*, which is defined as the energy–rich state of the interacting molecules at the top of the activation barrier

$$\underset{\textbf{Reactant}}{\textbf{A}} \rightleftharpoons \textbf{[transition state]} \rightleftharpoons \underset{\textbf{Product (s)}}{\textbf{P}}$$

The rate of a reaction is proportional to the concentration of the transition–state species. A rise in temperature because it increases thermal motion and energy, increases the number of molecules capable of entering the transition state; in many reactions, the reaction rate is approximately doubled by a 10°C rise in temperature.

Catalysts accelerate chemical reaction by lowering the free energy of activation. They combine with the reactants to produced a transition state having less free energy than the transition state of the uncatalyzed reaction. When the reaction products are formed, the free catalyst is regenerated.

Kinetics of enzyme–catalyzed reaction: the Michaclis–Menten Equation

The general principle of chemical reaction kinetics described above are also applicable to enzyme–catalyzed reactions. However, the latter show a distinctive feature not usually observed in non–enzymatic reaction, namely, the phenomenon of *saturation* with substrate. In Figure, we see the effect of substrate concentration on the rate of the enzymatic reaction A → P. At low substrate concentration, the reaction velocity *v* is proportional to the substrate concentration and the reaction is thus first– order with respect to the substrate. However, as the substrate concentration is increased, the reaction rate falls off and is no longer proportional to the substrate concentration; in its zone, the reaction is mixed–order. On further increase in substrate concentration, the rate becomes constant and independent of substrate concentration. In this range of substrate concentration, the reaction is zero– order with respect to the substrate; the enzyme is then saturated with its substrate. All enzymes show this saturation effect, but they vary widely with repect to the substrate concentration required to produce saturation

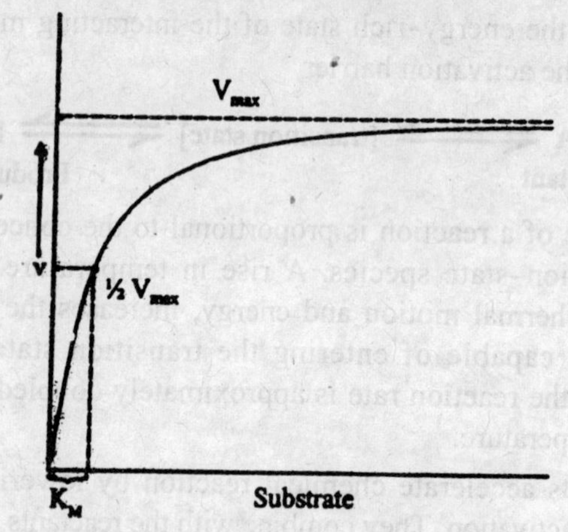

Figure. 5.24: ***Effect of substrate concentration on the rate of an enzyme-catalyzed reaction.***

The saturation effect led L. Michaelis and M.L.Menten in 1931 to a general theory of enzyme action and kinetics, which was later extended by G.E. Briggs and J.B.S.Haldane. According to this theory, which is basic to the quantitative analysis of all aspects of enzyme kinetics and inhibition, the enzyme ES, which then breaks down in a second step to form free enzyme and the product(s) P:

$$E + S \underset{k_2}{\overset{k_1}{\rightleftharpoons}} ES \qquad (1)$$

$$ES \underset{k_2}{\overset{k_1}{\rightleftharpoons}} E + P \qquad (2)$$

Both reactions are considered to be reversible; k_1, k_2, k_3 and k_4 are specific rate constant for the reactions designated.

In the following derivation of the Michaelis–Menten equation, which is that of Briggs and Haldane, [E] represents the total enzyme concentration (the sum of the free and combined enzyme),[ES] is the concentration of the enzyme–substrate complex, and [E]–[ES] represents the concentration of free, or uncombined, enzyme.[S] represents the substrate concentration, which is ordinarily for greater than[E], so that the amount of S bound by E at any given time is negligible compared with the total concentration of S. The derivation begins by considering the rates of formation and breakdown of ES. The rate of formation of ES from E + S given by

$$\frac{d[\mathrm{ES}]}{dt} = k_1\,([\mathrm{E}] - [\mathrm{ES}])[\mathrm{S}] \tag{3}$$

The rate of formation of ES from E + P is very small and may be neglected.

Similarly, the rate of breakdown of ES is given by

$$\frac{-d[\mathrm{ES}]}{dt} = k_2\,[\mathrm{ES}] + k_3\,[\mathrm{ES}] \tag{4}$$

When the rate of formation of ES is equal to its rate of breakdown, i.e., when the reaction system is in a steady state, with the ES concentration remaining constant,

$$k_1\,([\mathrm{E}] - [\mathrm{ES}])[\mathrm{S}] = k_2\,[\mathrm{ES}] + k_3\,[\mathrm{ES}] \tag{5}$$

Rearranging this expression, we have

$$\frac{[\mathrm{S}][\mathrm{E}] - [\mathrm{ES}]}{[\mathrm{ES}]} = \frac{k_2 + k_3}{k_1} = K_M \tag{6}$$

The "lumped" constant K_M which replaces the term $(k_2 - k_3) / k_1$, is called the *Michaelis–Menten constant.* From this equation, the steady–state concentration of the ES complex may be obtained by solving for [ES]:

$$[\mathrm{ES}] = \frac{[\mathrm{E}][\mathrm{S}]}{K_M + [\mathrm{S}]} \tag{7}$$

Since the initial rate v of an enzymatic reaction is proportional to the concentration of the ES complex, we can write

$$v = k_3\,[\mathrm{ES}] \qquad (8)$$

When the substrate concentration is so high that essentially all the enzyme in the system is present as the ES complex, we reach the *maximum velocity* V_{max} for which we can write

$$V_{max} = k_3\,[\mathrm{ES}] \qquad (9)$$

in which [E] is the total enzyme concentration.

We can now substitute for the term [ES] in equation(8) its value from equation(7):

$$v = k_3 \frac{[\mathrm{E}][\mathrm{S}]}{K_M + [\mathrm{S}]} \qquad (10)$$

TABLE: K_M FOR SOME ENZYMES

Enzyme and substrate	K_M *(mM)*
Catalase	25
H_2O_2	
Hexokinase	
Glucose	0.15
Fructose	1.5
Chymotrypsin	
N–Benzoyltyrosinamide	2.5
N–Formyltyrosinamide	12.0
N–Acetyltyrosinamide	32
Glycyltyrosinamide	122
Carbonic anhydrase	
HCO_3^-	9.0
Glutamate dehydrogenase	
Glutamate	0.12
*–Ketoglutarate	2.0
NH_4^-	57
NAD_{ox}	0.025
NAD_{red}	0.018

TABLE: EFFECT OF SUBSTRATE STRUCTURE ON VMAX FOR D–AMINO ACID OXIDASE (DATA ARE RELATIVE TO D–ALANINE = 100)

Substrate	Relative V_{max}
D-Tyrosine	297
D-Proline	231
D-Methionine	125
D-Alanine	100
D-Valine	55
D-Histidine	9.7
Glycine	0.0

If we now divide this equation by equation (9), we obtain

$$\frac{v}{V_{max}} = \frac{k_3 \dfrac{[E][S]}{K_M + [S]}}{k_3[E]} \qquad (11)$$

Solving for v,

$$v = \frac{V_{max}[S]}{K_M + [S]} \qquad (12)$$

This is the *Michaelis-Menten equation;* it defines the quantitative relationship between the enzyme reaction rate and the substrate concentration [S] if both V_{max} and K_M are known.

An important numerical relationship emerges from the Michaelis-Menten equation in the special case when $V = \frac{1}{2} V_{max}$ We then have

$$\frac{V_{max}}{2} = \frac{V_{max}[S]}{K_M + [S]}$$

If we divide by V_{max}, we obtain

$$\frac{1}{2} = \frac{[S]}{K_M + [S]}$$

Rearranging,

$$K_M + [S] = 2[S]$$

$$K_M = [S]$$

We may therefore conclude that K_M *is equal to the substrate concentration at which the velocity is half maximal.* K_M has the dimensions moles liter^{-1}.

Figure shows that K_M can be extrapolated graphically from data on the effect of substrate concentration on the reaction velocity; its magnitude is independent of enzyme concentration. Note that K_M is not a fixed value. It may vary with the structure of the substrate, with pH, and with temperature. In those enzymes having more than one substrate, each substrate has a characteristic K_M. Under intracellular conditions, enzymes are not necessarily saturated with their substrates.

The maximum velocity V_{max} varies widely from one enzyme to another. It also varies with the structure of the substrate, with pH, and with temperature.

The Michaelis constant, as we noted above, is an experimentally determined quantity, which in the idealized case is represented by

$$K_M = \frac{k_2 + k_3}{k_1} \quad (6)$$

However, in many enzymatic reactions, k_2 and k_1 may be very large compared with k_3. In such reactions, the rate-limiting step in the over all reaction is the slow step, ES $\xrightarrow{k_3}$ P. k_3 is then negligibly small, and equation (6) simplifies to the expression

$$K_M = \frac{k_2}{k_1}$$

Under these conditions, K_M is evidently the dissociation constant of the ES complex and is replaced by the expression K_s:

$$k_S = \frac{[E][S]}{[ES]}$$

Unfortunately, K_M and K_S are frequently used interchangeably. K_M should not be regarded as the dissociation constant of the ES complex unless specific information is available that k_3 is very low compared with k_2 and k_1.

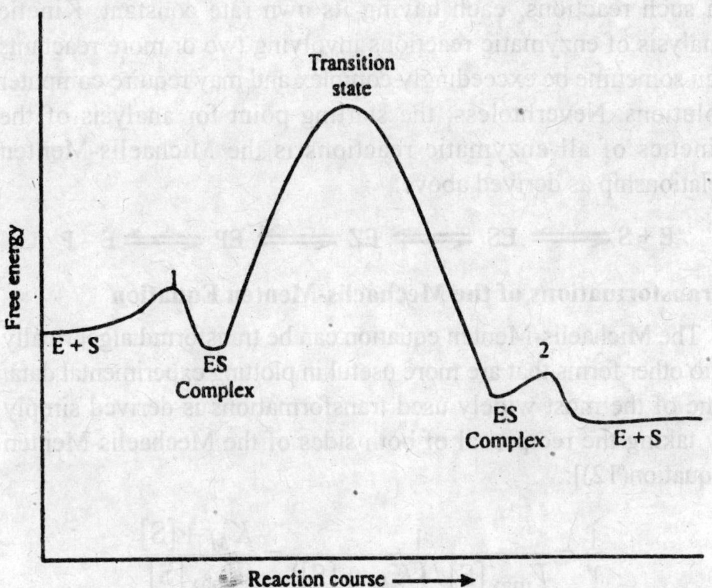

Figure. 5.25: ***Energy diagram for an enzyme-catalyzed reaction. Small energy barriers exist at points 1 and 2:***

The Michaelis-Menten equation is fundamental to all quantitative treatment of enzyme action. Its derivation from first principles, as shown above, leads to many other useful relationships. But it must be stressed that most enzymes show kinetic behaviour that is much more complex than the idealized case we have just treated. For one thing, our formulation assumed that there is but one enzyme-substrate complex. However, it now appears likely that most enzyme-catalyzed reactions involve two or three enzyme-substrate complexes, acting in the following sequence: in which EZ is the true transition-state complex and EP an enzyme-product complex. Furthermore, it must be pointed out that in most enzymatic reactions there is more than one substrate molecule and there may be two or more products. In a reaction with two substrates S_1 and S_2, there may be three enzyme-substrate complexes, namely, ES_1, ES_2, and ES_1S_2. If the reaction has two products P_1 and P_2, there may be at least three additional complexes EP_1, EP_2 and EP_1P_2. Many intermediate steps occur in such reactions, each having its own rate constant. Kinetic analysis of enzymatic reactions involving two or more reactants can sometime be exceedingly complex and may require computer solutions. Neverhteless, the starting point for analysis of the kinetics of all enzymatic reactions is the Michaelis-Menten relationship as derived above.

$$E + S \rightleftharpoons ES \rightleftharpoons EZ \rightleftharpoons EP \rightleftharpoons E - P$$

Transformations of the Mechaelis-Menten Equation

The Michaelis-Menten equation can be transformd algebrically into other forms that are more useful in plotting experimental data. One of the most widely used transformations is derived simply by taking the reciprocal of both sides of the Mechaelis-Menten [equation(12)]:

$$\frac{1}{v} = \frac{1}{V_{max}[S]/(K_M + [S])} = \frac{K_M + [S]}{V_{max}[S]}$$

Rearranging, we have

$$\frac{1}{v} = \frac{K_M}{V_{max}[S]} + \frac{[S]}{V_{max}[S]}$$

which reduces to

$$\frac{1}{v} = \frac{K_M}{V_{max}} \cdot \frac{1}{[S]} + \frac{1}{V_{max}} \tag{13}$$

Equation (13) is the *Lineweaver-Burk equation,* which represents a straight line with a slope of K_M/V_{max} and an intercept of $1/V_{max}$ on the 1/v axis. This line is obtained by plotting 1/ v vs. 1/[S]. Such a "double-reciprocal" plot has the advantage that V_{max} can be more accurately arrived at than from the simple plot *v vs.* [S]; in the latter, V_{max} is approached asymptotically and its value therefore is uncertain. The intercept on the abscissa of the

Lineweaver-Burk plot is $-1/K_M$. The Lineweaver-Burk plot can also give valuable information on enzyme inhibition, as we shall see below.

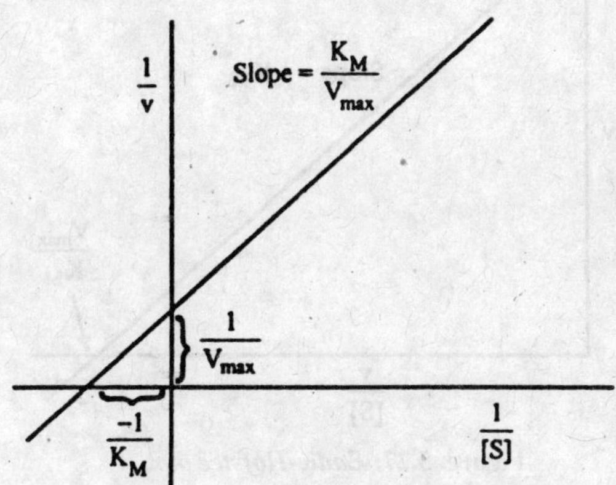

Figure. 5.26: ***Lineweaver-Burk plot.***

An other useful transformation of the Michaelis-Menten equation is obtained by multiplying both sides of equation (13) by V_{max} (v) and rearranging to yield the equation

$$v = -K_M\left(\frac{v}{[S]}\right) + V_{max}$$

When v is plotted against $v/[S]$, the plot shown in Figure 5.27 results. This plot (the *Eadie- Hofstee plot*) not only yields V_{max} and K_M in a very simple way but also magnifies departures from linearity which might not be seen in a Lineweaver-Burk plot.

Dependence of enzyme reaction rates on cofactores

The Michaelis-Menten equation not only defines the quantitative relationship between substrate concentration and enzyme reaction rate but can also be applied to the relationship between cofactor concentration and reaction rate, in the case of those enzymes requiring a cofactor for activity. The following type of equilibrium is assumed:

$$\underset{\text{(Inactive)}}{\text{E + coenzyme}} \rightleftharpoons \underset{\text{(Active)}}{\text{E – coenzyme}}$$

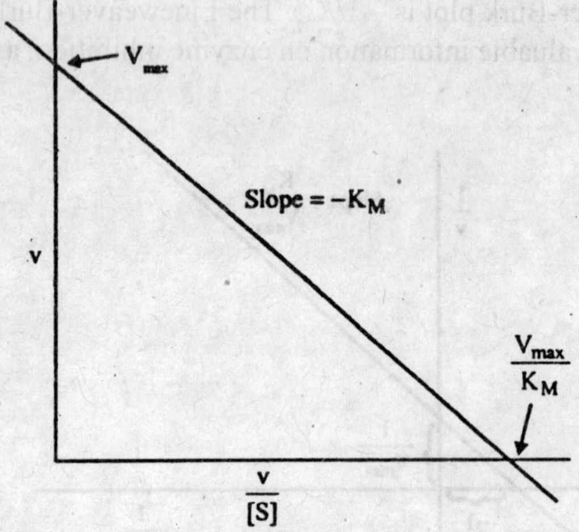

Figure. 5.27: ***Eadie-Hofstee plot.***

The enzyme-coenzyme complex may then bind the substrate to form the enzyme-coenzyme-substrate complex whose

$$E - \text{coenzyme} + S \rightleftharpoons S - E - \text{coenzyme}$$

concentration determines the overall reaction rate. It is evident from these equilibria that such an enzyme will show a saturation phenomenon not only with its substrate but also with its cofactor.

Just as we may have a Michaelis constant K_M for substrate affinity, defined as the concentration of substrate at which half-maximal velocity is observed, we may also have a Mechaelis constant to express affinity of the enzyme for its cofactor, similarly defined as the concentration of cofactor at which half-maximal activity of the enzyme is observed. To estimate the K_M for a cofactor, the substrate concentration is held constant at a saturating level and the effect of cofactor concentration on velocity is measured.

Table lists K_M values for the enzyme glutamate dehydrogenase, which requires the cofactor NAD to accept H atoms. The reaction catalyzed by this enzyme is

$$\text{Glutamate} + NAD_{ox} \rightleftharpoons \alpha\text{-ketoglutarate} + NAD_{red} + NH_3$$

The two substrates glutamate and α- ketoglutarate have

characteristic K_M values, as do the oxidized and reduced forms of NAD, the cofactor. In this and many similar enzymatic reactions involving coenzymes, the coenzyme participates in the reaction as though it were a substrate, and it can be treated by the same kinetic formalism.

6

Plasma Proteins in Normal and Abnormal Individuals

Inherited susceptibilities underlie the development of Psychiatric diseases such as Manic Depressive Psychosis (MDP) and Schizophrenia. Psychiatric geneticists have identified many families with multiple cases of mental illness and using the tools of Human Genome Project are trying to determine which genes are passed along in those families. However a larger sample size is required in order to obtain statistically significant results. Linkage analysis for these genetically complex disorders cannot be caried out without specifying the mode of inheritance and the degree of penetrance of the genes responsible for the disease. For neuropsychiatric disorders this information is not clearly known. The gene for manic-depressive psychosis was found close to the gene for colour blindness, on the long arm of X chromosome by Baron (1987), however it was reported to be at the opposite end of the X chromosome by Hodgkinson *et al* (1987). Other workers do not find evidence for X linkage of this gene. However the dominant gene at the tip of the short arm of chromosome 11 is also suggested to be responsible for MDP.

A susceptibility locus for schizophrenia was reported to be on chromosome 5. Again the replicate studies were unable to prove these findings. Evidence against any positive linkage of a Schizoprenia gene on chromosome 5 is also reported. Schizophrenia is arguably the worst disease affecting mankind, causing more then 1 in 100 people to be disabled early in their lives, and resulting in a disruption of society in some cases. Although the exact biochemical defect is not known but it is believed that the mesolimbic dopaminergic system is related to Schizophrenia, which is associated with premorbid deficits. Excess dopaminergic activity can cause dissociation of a person's drives and thoughts. The abnormal regulation of dopamine may be related to reduced frontal cortical inhibitory influences. The genes for glucocorticoid receptors are present on chromosome 5 q13 and disturbances in glucocorticoid metabolism can induce psychotic symptoms. Additional genes on other chromosome may also take part; thus the genetic factors underlying Schizophrenia may be heterogenous. It was found that there is a especial combination of psychological features in 56% of late onset schizophrenic patients. Mayer and colleagues (1993) reported that the schizophrenia beginning after age of 40 years is associated with less affective flattering and few thought disorders. And in the late onset type an increased frequency of depressive syndromes, peculiar delusions, olfactory hallucination was observed.

Neuroleptic, antipsychotic drugs revolutionized the treatment of schizophrenia by ameliorating schizophrenic symptoms. However only 30% of patients show a limited improvement and 10-20% show little or no improvement. Manic depressive psychosis (MDP) is a severe neuropsychiatric disorder, that can be unipolar or bipolar. It is characterized by abnormalities of mood and cognition as well as neurovegetative functions such as sleep, appetite and libido, and is estimated to affect 1% of the world's population. Although the mode of transmission is uncertain, family, twin and adoption studies strongly suggested a genetic involvement.

Psychoactive antidepressants and antistimulant drugs that help to ameliorate depression and mania are thought to act by affecting the catecholamine neurotransmitter system such as adrenaline, noradrenaline and dopamine. Noradrenaline secreting neurons are located in the brain stem, especially in the locus ceruleus. Defects in dopamine metabolism are also reported to be involved in the development of major mental disorders. Dopamine is involved in the synthesis of noradrenaline, so the alterations in the noradrenergic and dopaminergic neurotransmission may be indirectly implicated in the pathophysiology of MDP.

Establishing the role of genetic factors in the aetiology of mental illness has represented a formidable challenge. The separation of environment from intrinsic biological factors and the complexities of psychiatric diagnosis are major obstacles in this endeavor. Nevertheless evidence for biological and genetic contributions to the aetiology makes these major affective disorders excellent candidates to address this issue. The result of family, twin and adoption studies consistently indicate that genetic transmission is of major importance in the pathogenesis of this disease. As the biochemistry of MDP is not clear, linkage analysis is one of the best available method for unravelling the genetics and neurobiology of at least some forms of this complex disorder. Mutation affecting the tyrosine hydroxylase (TH) gene, which encodes the rate limiting enzyme for the synthesis of adrenaline, noradrenaline and dopamine, might be responsible for causing the MDP phenotype.

Addiction is not the failure of will or mortality, but a chronic disease. Addictive drugs serve as a reward; so called positive reinforcers in the language of behavioural psychology. Psychiatrists reported a high prevalence of antisocial behaviour, depression and paranoia in cocaine abusers. Biochemists reported unusually low level of certain messenger molecules or receptors in the brains of chronic alcoholics. One way of cutting through this cause and effect problem is to look for the genes responsible for the addictive behaviour. Addictive drugs like psychomotor stimulants (cocaine and amphetamines), depressants of central nervous system (ethanol) opiate narcotics (heroine and morphine) in part produce their rewarding effects of euphoria or pleasure through interaction with the mesolimbic dopamine system. This system involves the dopamine projections from the ventral tegmental area in the basal fore brain. The neural elements in this basal fore brain undergo adaptive changes during chronic drug exposure, these adaptive changes may reflect the dark side of drug dependence, that derived from the creation of a state of deficit or need. Cocaine, heroine, nicotine and alcohol all exert different patterns of influences on the brain's system of neurons and neurotransmitters. They have the ability to hijack some of the neural pathways in the brain. Drug increases the dopamine levels in certain parts of brain, mostly notably in a tiny cluster of nerve endings and neurons known as nucleus accumbens. Nicotine, cocaine, opiates and alcohol all reduce metabolic activity in brain by approximately 10-15%. Heroine stimulates opiate receptors in the brain. Cocaine acts most directly and potently on dopamine producing neurons. While alcohol in addition to stimulating dopamine release also directly makes neurons more sensitive to a sedating neurotransmitter known as GABA.

In the present study plasma protein concentrations of patients with schizophrenia, MDP and addiction were determined along with those in normal individuals. Electrophoretic plasma protein patterns have also been examined in the affected and normal individuals. This research was conducted with the following objectives:

1. To observe the difference in total plasma protein concentrations of normal individuals and patients suffering from three different types of psychological disorders at different age groups.
2. To observe and compare the qualitative separation of the plasma protein in patients with Schizophrenia, Manic Depressive Psychosis, Addiction and Normal individuals.

MATERIALS AND METHODS

Three types of psychiatric disorders were selected for the study. The case records of each and every patient were obtained from the hospital and all the medical, nursing therapy notes and all necessary information were studied. In almost every case a semi-standardized case summary was also available. In total 58 patients were selected; 29 suffering from SCHIZOPHRENIA, 19 suffering from MANIC DEPRESSIVE PSYCHOSIS (MDP), and 10 suffering from an ADDICTION. Normal individuals with no psychological symptoms were included as controls.

Collection of Samples

The blood samples were taken by vein puncture, using 3.8% sodium citrate as an anticoagulant. Plasma was separated by centrifugation and refrigerated.

Quantitative Estimation of Plasma Proteins

Total plasma proteins were estimated as described by Lowry (1951). From each blood sample 0.2 ml of plasma was used according to standard procedure. Absorbance was recorded at 750 nm against a reagent blank.

Qualitative Estimation of Plasma Proteins

The plasma was examined electrophoretically using the polyacrylamide gel electrophoretic (PAGE) technique as described by Maurer (1971) with minor modifications. 7.5% gel composition and Tris glycine buffer at 8.4 pH was used. Bromophenol blue in 4% sucrose solution was mixed with the plasma preparations then the samples were loaded under running buffer. A current of 3mA/tube was applied during PAGE. Gels were fixed in 12.5% TCA and stained in commassie brilliant blue R-250 stain. After destaining, the gels were photographed and zymograms were also prepared.

TABLE 6.1: THE CODE NUMBERS, AGES, AND PLASMA PROTEIN CONCENTRATIONS OF NORMAL INDIVIDUALS

Code no.	Age (yrs)	Plasma Protein conc. (µg/ml)
N1	23	1610
N2	23	1460
N3	23	1575
N4	23	1500
N5	22	1575
N6	45	1590
N7	35	1560
N8	42	1540

N9	23	1460
N10	37	1510
N11	22	1490

Mean age = 28.9, Range = 22-45 and Mean plasma protein concentration = 1533.6

Statistical Design

The experiment was designed as 4 × 2 factorial with a completely random design (CRD). Clinical condition was taken as factor A with 4 levels viz. Normal, Schizophrenia, Manic Depressive Psychosis (MDP), and Addiction. The age groups were taken as factor B with 2 levels viz. below 35 (16-34) and above 35 (35-66) yrs.

RESULTS

In this study 58 patients were selected and grouped into different categories according to their diagnostic status viz. Schizophrenia, MDP and Addiction. A code was given to each and very patient. A group of normal individuals was taken as control.

TABLE 6.2: THE CODE NUMBERS, AGES, AND PLASMA PROTEIN CONCENTRATIONS OF SCHIZOPHRENIC

Code No. (yrs)	Age (yrs)	Onset Age (µg/ml)	Plasma Protein Conc.
S1	29	17	975
S2	17	14	1260
S3	61	28	1525
S4	66	40	1510
S5	26	12	1510
S6	45	32	1490
S7	27	13	1425
S8	25	11	1500
S9	46	39	1425
S10	52	42	1400

S11	29	19	1540
S12	44	34	1460
S13	28	26	1490
S14	40	25	1525
S15	38	21	1475
S16	17	17	1500
S17	36	31	1500
S18	61	52	1390
S19	45	42	1460
S20	32	24	1510
S21	45	27	1590
S22	65	28	1460
S23	35	35	1500
S24	45	39	1500
S25	40	39	1625
S26	20	20	1475
S27	35	29	1525
S28	36	22	1425
S29	18	12	1540

Mean age = 38.03, Range = 17-66, Mean onset age = 27.24. Range = 11–52, Mean plasma protein concentration = 1465.86

TABLE 6.3: THE CODE NUMBERS, AGES, AND PLASMA PROTEIN CONCENTRATIONS OF PATIENTS WITH MANIC DEPRESSIVE PSYCHOSIS

Code No. (yrs)	Age. (yrs)	Onset age (µg/ml)	Plasma Protein Conc.
M1	27	26	1040
M2	41	35	1040
M3	52	37	1440
M4	32	29	1360
M5	16	15	1385

M6	55	55	1510
M7	30	30	1475
M8	30	23	1440
M9	55	55	1425
M10	35	—	1475
M11	39	21	1490
M12	40	39	1625
M13	20	20	1475
M14	35	29	1525
M15	36	22	1425
M16	42	38	1460
M17	45	25	1460
M18	28	26	1410
M19	19	17	1550

Mean age = 35.63, Range = 16–55, Mean onset age = 30.11, Range = 15–55 and Mean plasma protein concentration = 1421.6

TABLE 6.4: THE CODE NUMBERS, AGES, AND PLASMA PROTEIN CONCENTRATIONS OF ADDICT PATIENTS

Code No. (yrs)	Age. (yrs)	Onset age (μg/ml)	Plasma Protein Conc.
A1	39	35	1410
A2	27	21	1525
A3	24	13	1460
A4	22	19	1560
A5	35	24	1100
A6	18	12	1540
A7	42	38	1460
A8	45	25	1310
A9	28	26	1330

A10	19	17	1525

Mean age = 29.9, Range = 18-45, Mean onset age = 23, Range = 12-38 and Mean plasma protein concentration = 1421.0

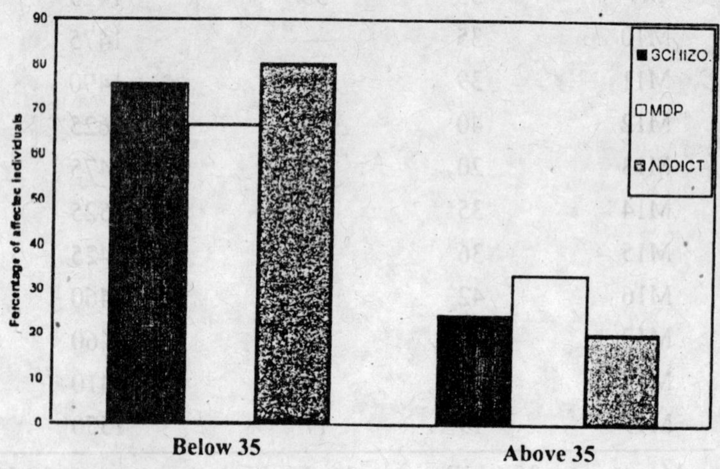

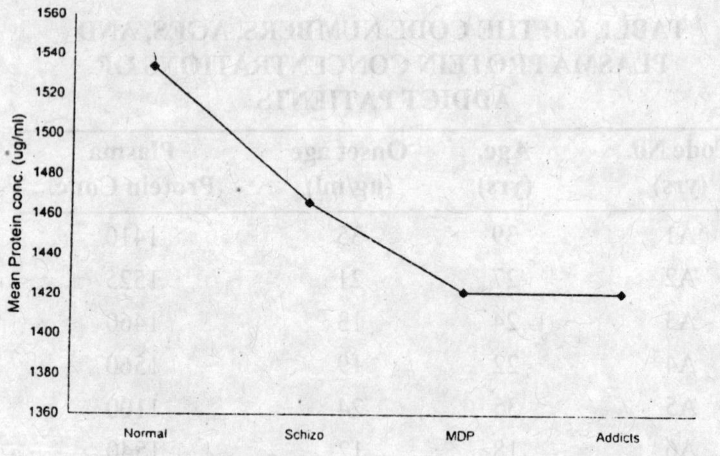

Figure 6.1: ***(a) Proportion of affected individuals in early and late onset ages. (b) Mean plasma protein conc. of Normal, Schizophrenics, MDP and Addicts.***

AGE AND DISEASE ONSET

The onset of disease was also noted from their case histories. The age of patients, Onset age distribution and plasma protein concentration for schizophrenics, patients with manic-depressive psychosis, and addicts are presented in Tables 6.2, 6.3, and 6.4 respectively. Table 6.1 represents the normal individuals.

TABLE 6.5: ANALYSIS OF VARIANCE OF TOTAL PLASMA PROTEIN CONCENTRATION OF NORMAL AND PATIENTS WITH PSYCHOLOGICAL DISORDERS

Sources of Variation	Degrees of freedom	Sum of Squares	Mean Square	F - Value
Treatment combination	7	815693.20	116527.60	6.493*
Factor A Clinical Condition	3	734077.60	244692.50	13.63*
Factor B Age Groups	1	51601.06	51601.06	$2.87^{n.s}$
Interaction A X B	3	30014.54	10004.84	$0.55^{n.s}$
Error	24	430717.80	17946.57	
Total	31	1246412.00		

$F_{.05}$ df [7] [24] = 2.42, df [3] [24] = 3.01, df [1] [24] = 4.26

The mean ages of onset for Schizophrenia, MDP and Addiction were 27.24 ± 10.6 st. dev., 30.11 ± 11.0 and 23 ± 8.1 st. dev. It was also noted that proportion of affected individual was much higher in early ages (below 35) than in the late ages (35 or above) as shown in the Figure (6.1a).

Quantitative Estimation of Plasma Protein Concentration

Mean plasma protein concentrations of normal individuals and of patients with Schizophrenia, Manic Depressive Psychosis and Addiction were 1533.6, 1465.8, 1421,6 and 1421.0 μg/ml as shown in Tables 6.1, 6.2, 6.3, 6.4 and 6.5 respectively and represented in Figure 6.1b.

Plasma protein concentrations were statistically analyzed by 4 × 2 factorial using Completely Random Design (CRD). The results are presented in Table 6.5. Clinical conditions were taken as factor A with 4 different levels viz. Normal, Schizophrenia, Manic Depressive Psychosis and Addiction. The age groups were taken as factor B with 2 levels, below 35yrs. and above 35yrs. of age.

TABLE 6.6: DUNCAN'S MULTIPLE RANGE TEST

1st	2nd	3rd
A – B = 67.8*	A – C = 112.1*	A – D = 112.6*
B – C = 44.3$^{n.s}$	B – B \ 44.8$^{n.s}$	
C – D = 0.5$^{n.s}$		

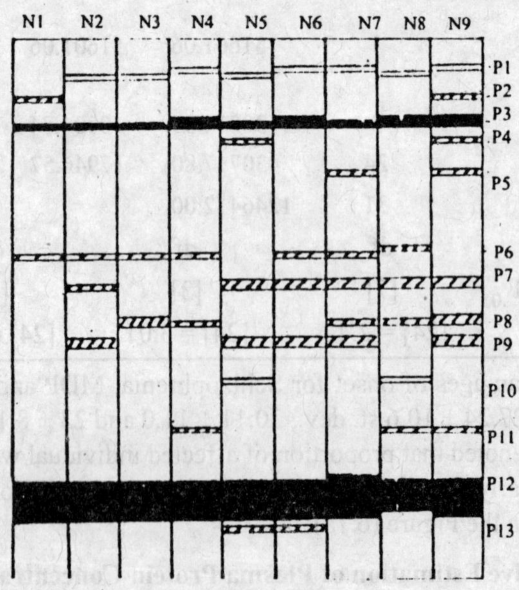

Figure 6.2: ***Zymogram showing qualitative separation of plasma proteins of normal individuals.***

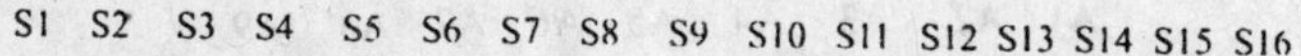

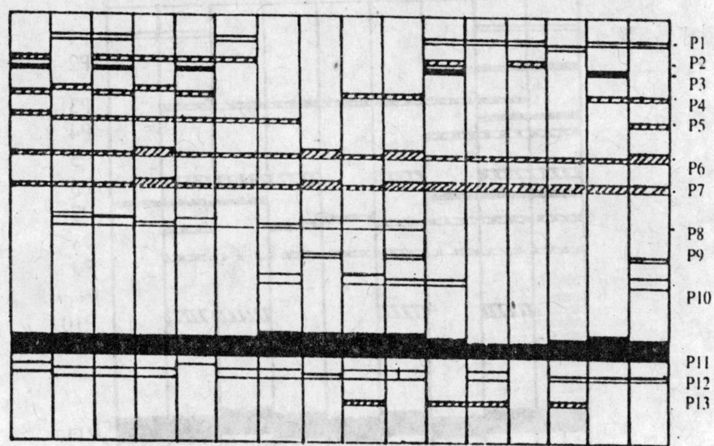

Figure 6.3: ***Zymogram showing qualitative separation of plasma proteins of schizophrenic patients.***

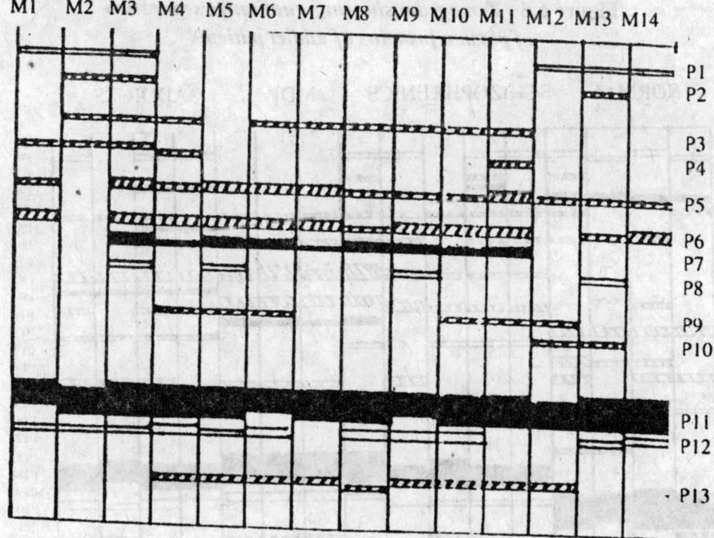

Figure 6.4: ***Zymogram showing qualitative separation of plasma proteins of MDP patients.***

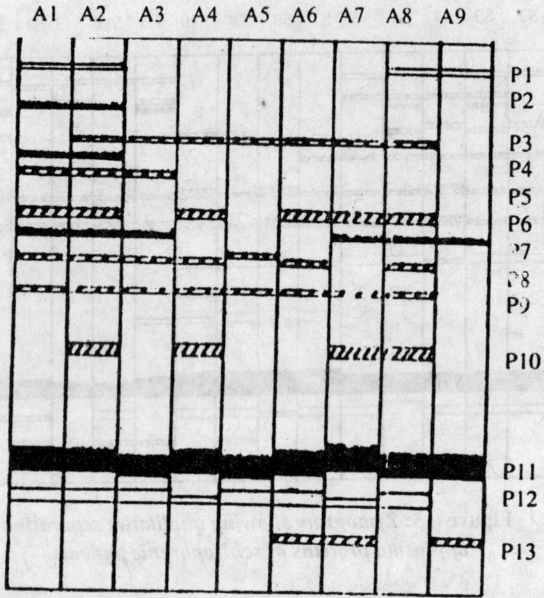

Figure 6.5 : ***Zymogram showing qualitative separation of plasma proteins of addict patients***

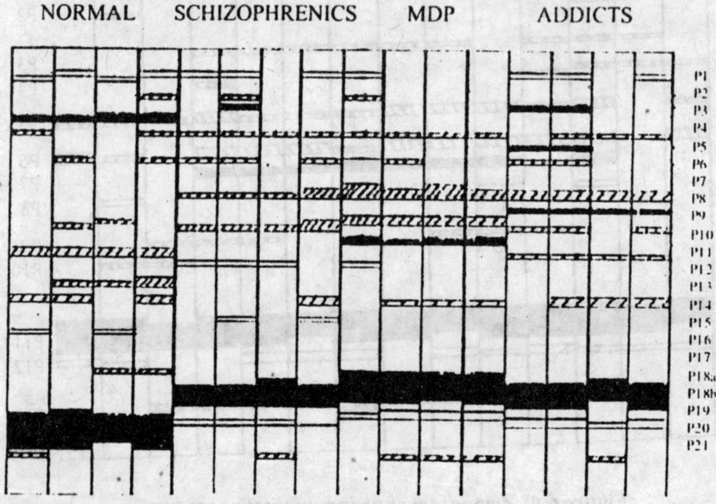

Figure 6.6: ***Zymogram showing qualitative separation of plasma proteins of normal individuals, schizophrenic, MDP and addict patients.***

Factor A was found to have significant effect on the plasma protein concentration. This means that the protein concentrations of normal, schizophrenics, manic depressive psychosis patients and addicts were significantly different. Factor B however had a non-significant effect indicating that the plasma protein concentrations in below 35 and above 35 age groups were not significantly different. The interaction between the two factors viz clinical conditions and age groups., was also non significant indicating that the levels of factor B viz. Age groups behaved consistently with the levels of factor A viz. Clinical conditions of individuals and vice versa.

Duncan's multiple range test indicates significant differences in the plasma protein concentration of normal individuals and all psychological disorders however, non-significant differences were found among these psychological disorders.

Qualitative Separation of Plasma Proteins of Normal, Schizophrenics, Manic Depressive Patients and Addicts

The plasma samples were also analyzed electrophoretically for normal, schizophrenics, patients with manic depressive psychosis and addiction by using polyacrylamide gel electrophoresis (PAGE). The zymograms of total protein bands obtained for normal, schizophrenics, MDP patients, and addicts are individually presented in Figures 6.2, 6.3, 6.4, and 6.5 respectively; these are compared in Figure 6.6.

Altogether 21 bands were observed of which 7 bands (P1, P5 P7, P10, P15 and P21) were common to all. Band P4, P14, P17, and P20 noted in normal plasma samples were missing from the plasma sample of all psychiatric patients, whereas a new band P8 was observed in the samples of all affected individuals with some variation in width. Band P2 observed in normal, schizophrenic and manic depressive psychosis samples was missing in the samples of addict individuals. While a dark band P3 and a very light and P19 were observed only in the plasma samples of Addicts and Schizophrenics. Like wise band P13 was only noted in the samples taken from schizophrenics and MDP affected individuals. Band P12 found in samples from normal individuals was also observed in the samples of addicts, but not in those from MDP and schizophrenics. The plasma samples of manic-depressive patients shown a very dark thick band P18a, while in the samples from schizophrenics and addicts P18b was present with same intensity and width, which is missing in the normal samples. Band P18 appeared in all the samples of psychologically affected individuals with slightly variable width and position. In the plasma protein samples from addicts two new bands, P6 and P9, also appeared. An unique band, P11 was observed only in the samples from patients with manic depressive psychosis. Similarly a light band P16 was observed in the plasma samples of schizophrenic patients.

DISCUSSION

In the present study the mean age of onset for schizophrenia, manic depressive psychosis and addiction was 27.2 ± 10.6 years, 30.11 ± 11.0 years, and 23 ± 8.1 years respectively. It has previously been reported that Schizophrenic patients come to the attention of doctors between the ages of 18 and 30 years. In addition Castle (1993) reported that the mean age for onset for schizophrenia was 31.2 yrs while Mayer *et al.*, (1993) reported it to be 34.2 ± 13.8 years. For MDP the mean onset age was reported to be 22 years and usually between 15 and 35 years. For drug addiction, the mean age of onset was reported to be 25.04 years by Aftab (1990) and between 20-30 years by Jamshed (1992). According to the national survey of drug abuse in Pakistan (1986), the average age of drug abusers in Pakistan is 24 years. However a higher proportion of drug addicts were found in younger age groups by Varma and Sharma (1993). In the present study the majority of these psychological disorders occurred in early ages i.e. between 16-35yrs. It was also noted from the case histories that some patients had close relatives with other psychological disorders, indicating that schizophrenia, MDP and addiction tend to run in the families of affected individuals i..e these are familial disorders. The risk of becoming schizophrenic in all first degree relatives was 13 times greater than in relatives of controls. Psychiatric morbidity was present in 34.8% of first-degree relatives of schizophrenic patients. Bipolar affective disorders were also observed more commonly in first degree relatives of patients. The risk has been estimated to be about 10% in all first-degree relatives of a schizophrenic patients and 50% for identical twins. The risk is higher in individuals with two schizophrenic parents compared to individuals with only one affected parent.

Quantitative Estimation of Total Proteins in Mentally Ill Patients and Normal

The result obtained from the present study showed significant differences in the total plasma protein concentration of psychologically ill persons and normal individuals. This indicates that there is a significant effect of these psychological conditions on total plasma proteins that may be due to a long course of antipsychotic drugs. Jamshed (1992) also reported that the values of plasma protein concentration were below normal in majority of cases. Low serum cholesterol concentrations may also contribute to disturbed behaviour reported an overlapping of the manic depressive psychosis and schizophrenia disorders. Paranoid personality disorder and drug addiction are biologically related to schizophrenia and within the boundaries of the schizophrenic spectrum. There is a familial relationship between the schizophrenia and other psychiatric disorders and in the relatives of schizophrenia other psychological disorders like alcoholism are very common. In the present study the interaction between clinical conditions and age groups was nonsignificant indicating that age groups behaved consistently in different clinical conditions for plasma protein concentration. Also observed nonsignificant relationship between age and epidemiology of schizophrenia.

Qualitative Analysis of Total Plasma Proteins in Normal and Patient with Phycological Disorders

In the present study it was observed that some protein bands (P4, P14 and P17) were missing from the sample of mentally ill patients. A few new bands were observed in the samples from schizophrenic- (band p16), manic-depressive psychosis- (band P11) and in addiction-patients (P6 and P9). In all samples of psychologically disordered individuals a thick very dark band made its appearance. The results indicate that the aetiology of these disorders (namely schizophrenia, MDP and addiction) is strongly related to disturbed levels of different plasma proteins. The variation in the number of protein bands could also be due to the action of anti psychotic/neuroleptic drugs that were administrated for prolonged periods to overcome the severity of illness. Studies of plasma cortisol during and after depression showed that the disorder is associated with disrupted cortisol production. Plasma cortisol levels were also found to be elevated in samples from MDP patients. By infusing 2-deoxyglucose, which inhibits intracellular glucose metabolism reported significantly higher plasma homovanillic acid (HVA) levels in schizophrenic patients. The deprivation of glucose has strong effect on central peripheral dopamine function, including an increase dopamine turn over. Alterations in nonadrenergic and dopaminergic neurotransmission have been indirectly implicated in the pathophysiology of MDP. Researchers have focused their candidate gene effort on loci encoding proteins involved in the function and regulation of these neurotransmitter system. A number of candidate genes reside in the distal region of 5q, including the D1 dopamin receptor gene DRDI (5q 35.1), the alpha and beta–Adrenergic receptor genes (5q 32-34), the alpha 1 subunit gene of the GABAa receptor (5q 34-35) and one of the glutamate receptor gene (5q 32-33). A partial trisomy on chromosome 5 was reported in two patients with schizophrenia in a small multiplex family. Manic depressive psychosis might be traceable to a single gene on the short arm of chromosome 11. Schizophrenia in several Icelandic families is co-inherited with a particular region of chromosome 5 reported that the same region of chromosome 5 is genetically unrelated to the incidence of schizophrenia, in a larger Swedish family. Genetic linkage between X-chromosome markers and bipolar affective illness was also reported. However Gershon *et al.* (1979) have presented evidence against close linkage. It is not surprising that a defect in a single gene may cause a complex and variable disorder in human behaviour or that the predisposition is apparently attributed to different families. Manic depressive psychosis and schizophrenia are heterogenous disorders, mutations at different loci are responsible for psychologically disordered phenotype.

Further work is required to relate the patient's plasma protein levels with the level of enzymes involved in the regulation of different neurotransmitters. Neurotransmitters such as dopamine, adrenaline, noradrenaline and serotonin have very active roles in determining the pathophysiology of individuals. The genetics of complex disease such as affective mental illness, which results from the interaction of several genes with environmental factors, can be determined in the following ways. (1) Studying the distribution of alleles of genetic markers that are close to the candidate genes in a large population of unrelated individual patients and control (association studies) (2) detecting the cosegregation of the disease and marker alleles within families through linkage studies.

The genetic markers whose gene products may possibly be supposed to play a role in the pathogenesis of affective mental illness are only few, such as the genes encoding neuroreceptors and enzymes involved in the biosynthesis and metabolism of neurotransmitters. The candidate gene approach which is used in both linkage and association studies is powerful and a number of new molecular techniques allow us to focus directly upon variations in DNA sequences that may be of functional significance.

7

Recombinant Proteins

An important issue in the design of production schemes for recombinant proteins is to minimize the number of recovery steps during the purification. This can be achieved by introducing highly selective unit operations either with or without the use of affinity "tags". Another issue of great importance is to develop improved methods for protein folding based upon the understanding of the mechanisms which govern folding, both *in vivo* and *in vitro*. The pathway of folding *in vivo* is dependent on the host and the growth conditions. Protein folding *in vitro* is affected by the solvent environment and other physical conditions, such as temperature. This chapter will describe different techniques for purificantion and folding of recombinant proteins expressed in *Escherichia coli*.

PROTEIN FOLDING

The Protein Folding Problem

In protein chemistry, significant efforts have been made trying to solve how information from the amino acid sequence is transferred into a three dimensional structure. The understanding of the protein folding problem has increased in recent years. From the beginning the topic was primarily of academic interest but has now become an issue of industrial importance. Production of recombinant proteins in heterologous hosts often lead to unfolded and inactive proteins, in particular when the production levels are high. In order to generate useful protein products it is necessary to refold the protein *in vitro*. The folding problem was elegantly outlined by Levinthal and is referred to as "the Levinthals paradox": Assume a protein with 100 amino acids where every residue can have two different conformations. Despite this very simplified model, the protein can adopt 10^{30} different conformations. Even if the protein is able to sample new configurations at the rate of 10^{13} per second it well need 3×10^9 years to fold, similar to the age of earth. The true allowed number of conformations for an amino acid is of course much higher than two, and the time for a protein to fold completely random to search for the native conformation must exceed life of the earth. In contrast to these statistical figures, folding of an α-helix is estimated to take place in a time scale of 10^{-6} seconds. Consequently, there must be a non-random pathway for a protein to fold into an active structure.

Normally, folding of a protein can be achieved without any extrinsic factors, as first described by Anfinsen (1973). Many different hypothetical models have been proposed for the folding process. It has been compared with a jigsaw puzzle. A nucleation model has also been proposed where small "nucleus" form randomly and these serve as templates upon which folding can proceed. The random formation of the "nucleus" is supposed to be the rate limiting step. An alternative hypothesis is the frame work model which is the classical way to look upon flying. Secondary structure elements are supposed to fold first. The tertiary structure is then being formed by packing the secondary elements together. This hypothesis presumes a certain stability of secondary structure elements. In the diffusion-collision-adhesion model, micro domains fold up and interact with each other to form a stable three dimensional structure. The rate constant for folding depends on the diffusion rate and the stability of certain structures. This model is similar to the frame work model but it does not assume partly already existing secondary structures. Another theory of protein folding proposes that proteins undergo a rapid nonspecific hydrophobic collapse at the start of the folding process in order to reach a state close to the molten globule. The hydrophobic effect is thought to be the driving force. As the protein becomes more compact, the number of possible interactions within the protein diminishes. Constraining the polypeptide backbone might increase the probability to reach the native state. From available experimental data it is not possible to build a simple kinetic folding model that is applicable to all studied proteins. When studying proteins with proline residues or disulfide bonds, the rate determining step of the folding process is often that of proline isomerization or disulfide bond rearrangement. The folding of larger multidomain proteins is even more unpredictable. In many different folding systems it has been found that the rate limiting step is late in the folding process. The rate determining step was concluded to go from a compact intermediate to the final packed native conformation. This intermediate was designated "the molten globule" (MG). There is confusion in the literature about the definition of the molten globule state. One of the main sources for confusion is that the term is used both for the original definition of a molten globule is a protein with a high secondary structure content but an absence of unique tight packing of side chains. It is very difficult to study the molten globule state, since the real intermediate molecule is unstable and transient. Therefore, the most frequently studied variant is a partial denatured protein. Even though this is not a true molten globule, it is closely similar.

PROTEIN FOLDING IN THE CELL

During *in vitro* refolding proteins aggregate in much lower concentrations than usually found *in vivo*. In biological systems there are several molecules helping newly synthesized proteins to fold. Molecular chaperones and foldases seem to prevent proteins from incorrect folding and aggregation and increase the rate of productive folding. One important function of chaperones is to reduce the concentration of misfolded and unfolded proteins in the cell and thereby decrease the aggregation rate. Some proteins are initially synthesized as precursors with prosequences that are removed after folding. BPTI, subtilisin and α-lytic protease are examples of such proteins. Prosequences have been demonstrated to affect folding at the kinetic level.

Although the presence of thiol/disulfide redox buffers significantly increases the rate of oxidative folding *in vitro*, it still takes place on a time-scale ranging from minutes to hours. If proteins are to fold faster *in vivo* the reaction must be catalyzed. There are a number of characterized proteins that catalyze disulfide formation *in vivo*. One of those is Protein Disulfide Isomerase (PDI). PDI is resident in the lumen of the endoplasmic reticulum (ER). It has two disulfide active resident in the lumen of the endoplasmic reticulum (ER). It has two disulfide active sites, one near the N-terminus and the other near the C-terminus. PDI catalyzes thiol-disulfide exchange by oxidation and reduction and thereby assists in rearranging protein disulfides. These proteins that catalyze disulfide formation are thought not to alter the pathway, but accelerate it. PDI is relatively non-specific, but has slightly higher specificity for cysteine containing stretches. In *E. coli*, several proteins involved in the formation of disulfide bonds have recently been discovered. The periplasmic protein DsbA, discovered in 1991 by Bardwel *et al.*, is assumed to serve as an oxidant of secreted proteins. The inner membrane protein DsbB is thought to transfer oxidizing equivalents from the cytoplasm to DsbA. A more efficient catalyst of disulphide rearrangement than DsbA is DsbC, a soluble protein found in the periplasm. Also, a fourth disulphide isomerase-like protein in *E. coli* has been identified as an inner membranes protein suggested to supply the periplasm with reducing equivalents. The isomerization of the proline imide bonds is often a rate limiting step in protein folding. Proteins with prolyl peptide isomerase activity have been found in most organisms and in different cellular compartment. *In vitro,* they have been shown to catalyze several slow folding reactions.

The high concentration of unfolded and partly folded proteins in the cell should without the presence of molecular chaperones result in aggregation and loss of a large fraction of proteins. Molecular chaperones protect against aggregation by binding and releasing proteins along their pathway to the native structure. They are found in all cell types and are divided into several families of proteins that are structurally unrelated. At first, they were denoted stress or heat shock proteins, since several of them were induced under heat or harsh conditions, although they function also under normal conditions. Molecular chaperones important to protein folding can at present be divided into two main families; the hsp 70 family which binds to the nascent polypeptide chain to prevent premature folding, and the hsp60 family or chaperonins which at a later stages promotes the correct folding of the protein. Several investigations have been made to estimate the fraction of proteins that need chaperonins to fold. A genetic study indicated that a minimum of 30% of the soluble proteins in E. coli requires chaperonins to reach the native state. Two other investigations base their calculations on the *in vitro* rate of GroEL action and the rate of protein synthesis in *E. coli.,* and reach values of 5% and 7.5%. It is however questionable to what degree data from *in vitro* experiments can be used when describing the *in vivo* action of chaperonins.

INCLUSION BODIES

Since prokaryotic organisms do not have the necessary machinery for complete processing and folding of eukaryotic proteins, production of eukaryotic proteins in prokaryotic organisms often results in formation of inactive proteins. One consequence of the genetic modification of microorganisms in order to produce recombinant proteins, is accumulation of insoluble intracellular aggregates, inclusion bodies. The aggregation of foreign recombinant proteins has been observed in most cases but high level expression of wild-type protein will under certain circumstances also lead to inclusion body formation. Several factors influence the tendency of recombinant proteins to misfold or aggregate. The recombinant protein use to be produced in a heterologous host or expressed in levels that by far exceed the normally expressed amounts in the organism. Therefore, the environment is often significantly different in both chemical properties and in the availability of "helping proteins" e.g. foldases and chaperones. The formation of inclusion bodies is determined by the rates of two parallel competing reactions, the first order folding and the second order aggregation. Hence, the rate of aggregation is dependent on the rate of translation within the cell and also on the rate of folding for the protein of interest. The insolubility results from the formation of incorrectly or partly folded proteins. Attempts have been made to elucidate if there are particular properties of a protein that are connected with the inclusion body formation, but mostly without success. Neither the propensity to form precipitate, the size of the expressed protein nor the hydrophobicity are shown to be correlated to the inclusion body formation. Experiences suggest that the host cell physiology is the most important factor in inducing inclusion body formation. Hence, under different fermentation conditions the same protein might be able to form inclusion bodies or stay in the solution. Analysis of how amino acid substitutions affects the inclusion body formation indicates that the folding pathway is more important than the stability of the fully folded protein itself. The effect of other physical conditions, such as temperature and pH, may also be due to an alteration of the folding pathway. There are however, different methods to influence the inclusion body formation. One way is to make point mutations in the particular protein, to increase the solubility during folding. Large deletions or insertions can also affect the propensity to form inclusion bodies. Solubilizing linkers can affect the inclusion body formation, as shown for thioredoxin by LaVallie and co-workers (1993). Alternatively, one can over express chaperones or other proteins that can help in folding and thereby increase the solubility of the target protein. To obtain active proteins from intracellular aggregates, they are to be resolved and renaturated.

How to obtain functional protein in adequate quantities for research or commercial purposes is a main issue in both industry and academic. Here, the problems are often related to the production of functional proteins in a foreign host or, if this is not possible, how to perform *in vitro* refolding successfully. The main challenge is to overcome aggregation of unfolded or partially folded intermediates, which compete with the correct folding pathway. Since aggregation is a second-order process and correct folding the result of first order reactions, the concentration of the protein to fold is a critical factor. To obtain a functional protein one can either refold the protein produced as inclusion bodies *in vitro* or avoid the formation of inclusion bodies. Some proteins

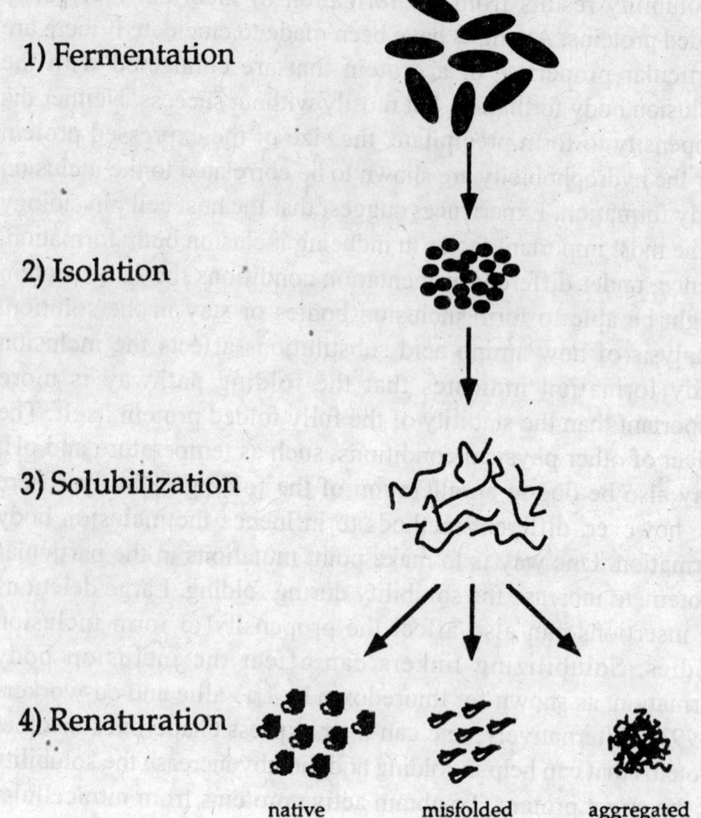

Figure 7.1: ***Strategy for production and recovery of proteins that form inclusion bodies. 1) The extent of inclusion body formation is dependent on the conditions during fermentation, such as temperature and chemical environment. Also the rate of protein synthesis is affecting the solubility of the proteins within the cell. 2) Isolation of the inclusion bodies starts by disrupting the cell membrane proteins, washing with a low concentration of denaturing chemicals and thereafter centrifugation is recommendable. 3) To solubilize the inclusion bodies, high concentration of denaturing agents are usually needed. When the target protein contains cysteines, a reducing agent has to be added to prevent non-native disulfide formation.4) The renaturation step is normally performed using stepwise dilution or dialysation. Composition of the solution used during renaturation is dependent on the protein of interest and the conditions have to be experimentally determined for each product. If native disulfides have to form, a low concentration of a reducing agent, such as β-mercaptoethanol, can make the refolding process more efficient.***

are very sensitive to degradation or deleterious to the host, and therefore it might be an advantage to produce them as inclusion bodies. Fusion to the *E. coli* TrpE protein can for example increase the propensity of a protein to form inclusion bodies. Other proteins are very hard to refold *in vitro* into an active form and are therefore preferably produced as soluble and native in the host cell.

Isolation and Solubilization of Inclusion Bodies

Since the inclusion bodies are intracellular or periplasmic structures, cells have to be fractured. The cells can be distributed through high pressure homogenization or sonication and the inclusion bodies can be isolated. As the membranes contain a number of proteolytic enzymes it is of importance to extract these contaminants or inhibit their activity. The most generally applicable method for selective extraction is washing the inclusion bodies with a buffer containing a chelating agent (EDTA), a neutral detergent (Triton X-100) and a chaotrop (Urea).

When the inclusion bodies are isolated they are to be solubilized . The forces involved in inclusion body formation are mainly non covalent and the presence of disulfides do not play any major role, even though disulfides in inclusion bodies have been reported. Solubilization, in most cases, involves total denaturation of the protein. If the protein has substantial amount of native conformation in the inclusion bodies it is possible to choose a weaker solubilizing buffer that do not disrupt the already formed structure. The most commonly used solubilizing agents are chaotropic salts or detergents. Once solubilized, the thiols are able to form intra or intermolecular disulfides. Therefore, its essential to include a reducing agent in the solubilizing mixture, in order to keep the thiols in the reduced state, or protect them by forming S-derivatives as have been used in the production of insulin.

RENATURATION

Thereafter, the protein has to be refolded *in vitro*. In some cases it is necessary to purify the protein before the renaturation step. DNA and host proteins might interfere with the refolding step by interaction with partly folded molecules. If purification in denaturating conditions is needed, gene fusion technology is useful. By fusing the protein of interest with an affinity tag that is usable under non-native conditions, the purification step can be both easy and selective. Affinity handles that are frequent used are the poly-histidine tails. The charged histidines residues are used because of their ability to bind to metal ions. It has been shown that using six histidines in the tail makes it possible to purify in denaturating conditions (6M guanidine hydrochloride). However, the purification method requires the imidazole residues in the non-protonated state. Poly-arginine tails have also been shown to be usable in denaturating conditions since their ability to bind to anionic residues are independent of the three dimensional structure.

One of the main issues in refolding proteins is to inhibit aggregation. The low solubility of unfolded proteins necessitates very dilute protein solutions in the refolding step. An alternative in order to reduce the reaggregation is dialysis or step wise dilution of the protein mixture. The staged dilution approach is critically dependent on the particular protein. Therefore the conditions have to be optimized carefully by experimentation for each protein. It is also possible to increase the solubility of a protein by chemical modification. This strategy could be used either to prevent covalent aggregation by disulfide formation or to change the net-charge of the protein by modifying the amides. Buffer additives which reduces the aggregation are also usable, especially when the particular protein needs a cofactor or a metal ion to achieve the native state. Other frequently used additives are polyethylene glycol, lauryl maltoside and arginine. Proteins can also be rendered either less or more soluble by gene fusion technology. There are several protein fusions that have proven to increase the solubility of the product. One example of a fusion protein that can affect the solubility is glutathione-S-transferase. Also, by using the highly soluble protein A domain (Z) as a solubilizer of misfolded and multimeric IGF-I, Samuelsson *et al.* were able to show that high yields of native protein could be recovered. The aggregation and precipitation of misfolded material was reduced. Since the refolding could be performed in high concentrations without any expensive chemical, the approach is well suited for large scale refolding. This method has also been utilized for IGF-II with success.

Creating micelles with the unfolded protein inside has also been tried in order to avoid aggregation. Another way to prevent aggregation in the refolding step has been elegantly shown by Rudolph and co-workers. They have fused α-glucosidase to a hexa-arginine polypeptide. This construction allows them to attach the fusion protein to a solid support containing polyanionic groups. Upon removal of the denaturant, α-glucosidase are able to refold without aggregation.

DISULFIDE FORMATION

Since many of the proteins of interest in biotechnology are extracellular and use disulfides for stabilization, disulfide formation is of great importance. These problems have motivated many researchers to study the periplasmic space of *E. coli* with the regard to disulfide formation. Formation of a disulfide is a two electron oxidation reaction and requires an oxidant, for example oxygen with metal ions, oxidized DTT or oxidized β-metcaptoethanol. The most abundant disulfide exchange agent in most cells is glutathione. The reaction starts with an attack of a nucleophilic thiolate ion on the disulfide bond. This reaction gives rise to a symmetrical transition state, with the charge uniformly distributed on the two sulfurs. The rate of disulfide formation is high with a basic attacking group, and a more acidic leaving group. The leaving group will be protonated. Therefore, increasing the pH will increase the rate of disulfide formation until the attacking group is mainly in the thiolate ion form. Disulfide formation can proceed by one of two chemical processes. The cheapest option for disulfide formation is air oxidation catalyses by metal ions . The reaction is irreversible and therefore most effective when the protein can adopt the native conformation before the disulfides have formed. If the protein is unable to form such stable intermediates, some of the molecules may get kinetically trapped with thermodynamically unstable, mispaired disulfides. Therefore,

a) air oxidation, catalyzed by metal ions, irreversible

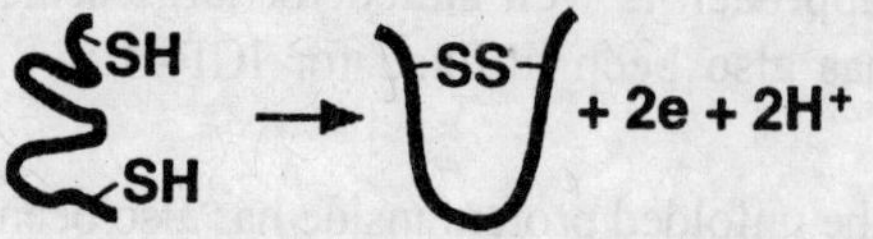

SH
SH → SS + 2e + 2H⁺

b) disulfide exchange, reversible

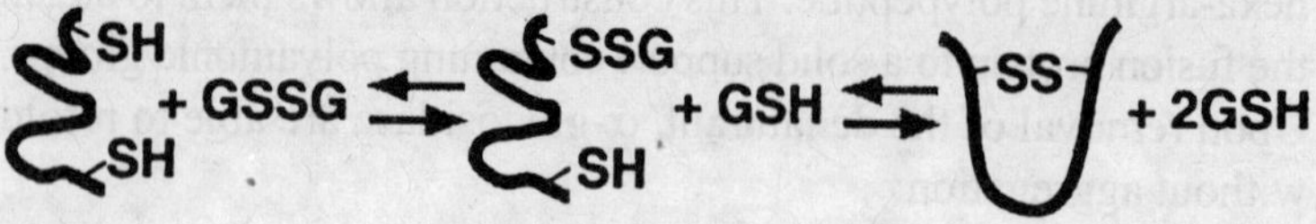

SH
SH + GSSG ⇄ SSG
SH + GSH ⇄ SS + 2GSH

Figure 7.2: (a) ***Disulfide formation with oxygen as the oxidant. This is an irreversible reaction and therefore molecules might get kinetically trapped with nonnative disulfides. If disulfide formation with oxygen as the oxidation agent is going to be successful, the protein must be able to form the native structure without having the native disulfides. One way to solve this problem, is to add small amounts of a reducing agent in to the refolding buffer form. Here the protein is able to break and reform disulfrides until the thermodynamically most stable ones are formed. (b) A disulphide exchange reaction with glutathione. Since this reaction is rerersible, mispaired disulfides can be reduced and new disulfides are able to***

air oxidation may be more successful if small amounts of a reduced thiol compound, such as β-mercaptoethanol is present. *In vivo* disulfide formation of a protein is a reversible process in the ER with glutathione as a redox buffer. The thermodynamic stability for protein disulfides varies greatly, and is dependent on the structure of the protein and the number of amino acids between the cysteines. According to Zhang *et al*. odd numbers of intervening amino acids unfavour disulfides compared to even numbers. To form a disulfide between two adjacent cysteines a significant amount of energy is needed. If there are long stretches with amino acids between the cysteines, the disulfide formation is more dependent on the 3D-structure of the protein and is therefore more unpredictable. Charges around the cysteines can also affect the disulfide formation. Positive charges stabilize the reduced state, consequently retaining free cysteines. Negative charges around the cysteines stabilize the oxidized state, promoting disulfides formation. pKa for a cysteine is approximately 8.5 but varies in different proteins, depending on the three-dimensional environment.

ENGINEERING OF PROTEIN FOLDING

Attempts to get a more efficient refolding have been made by adding chaperones and foldases to the folding mixture. Since aggregation during the refolding step is due to a slow folding pathway, it might be successful to add foldases to the refolding mixture. Such experiments have been tried with both disulfide and prolyl isomerases. Lilie *et al.* have shown that refolding is accelerated when prolyl isomerase is used in the refolding of an antibody. Although the isomerase is shown to act as a true catalyst it does not affect the aggregation since prolyl isomerization is a late step in the folding process. On the other hand, when using PDI as a catalyst for disulfide formation, researchers have been able to improve the yields of native protein remarkably. Because of the high amounts of "helping proteins" needed it is not yet feasible to use them *in vitro*. Another approach to favour correct folding is to alter the environment in host cell be co-expression of mammalian chaperones and catalysts. However, the results are unpredictable and the effects often minor, as in the antibody fragments expressed in *E. coli*.

The use of selective binders for assisted folding, when the particular protein lacks the necessary conformational information to fold completely, has been successfully tried. Monoclonal antibodies have been used for *in vitro* refolding of protein S, immobilized trypsin for refolding of Bowman-Birk type proteinase inhibitors and also simultaneous expression of two receptor partners that are able to form a heterodimer. Samuelsson and co-workers showed that co-expression of a specific binding protein in *Escherichia coli* can significantly improve the relative yields of correctly folded human insulin growth factor I (IGF-I). The molecule has been shown to possess a thermodynamic folding problem and is unable to quantitatively form the native disulfides.

a) Co-expression in *E. coli*

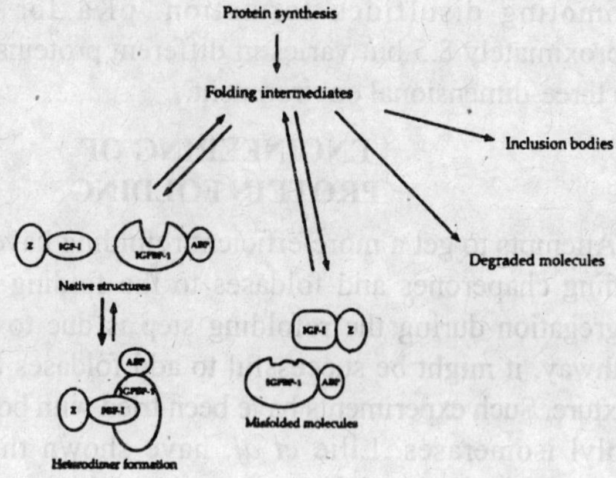

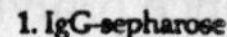

b) Co-purification

1. IgG-sepharose

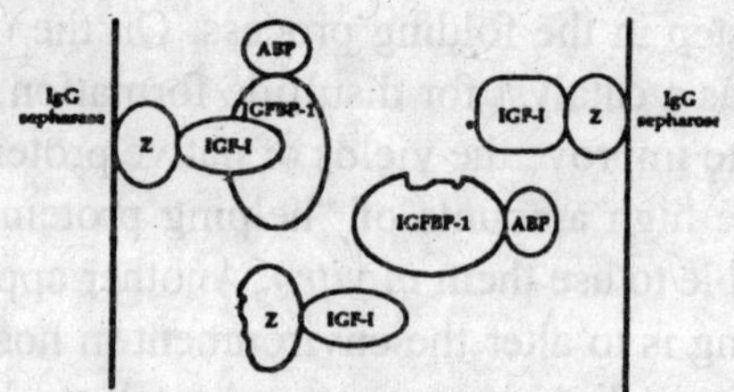

2. HSA-sepharose

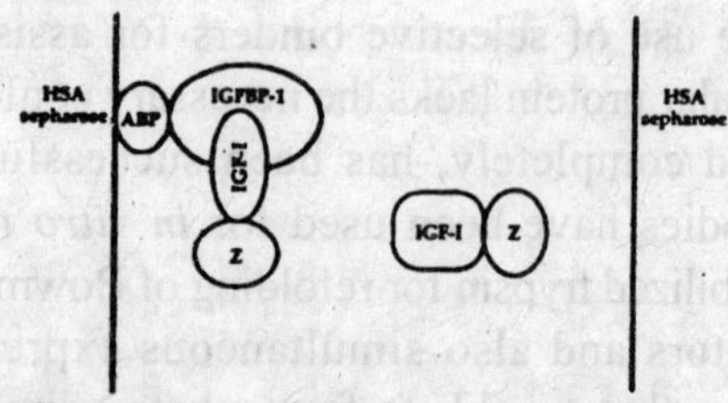

Figure 7.3: **a)** ***Affinity assisted*** in vivo ***folding by co-expression in*** Escherichia ***coli. The binding protein (IGFBPI) and native IGF-I form a complex and thereby lower the apparent concentration of the native conformation. The equilibrium between misfolded and native IGF-I is shifted towards the native form. b) Co-purification of IGF-I and IGFBP-I. By using different affinity tags on the target proteins, two separate and sequentially used affinity columns are able allow enrichment of the correctly folded fullength material.***

Instead, all attempts to produce IGF-I in heterologous hosts have resulted in misfolded species. Yet, when the protein was allowed to fold in the presence of its binding protein (IGFBP-I) it was able to quantitatively attain the native conformation. Therefore, IGF-I turned out to be an interesting molecule for affinity-assisted folding. Since both the binding protein and IGF-I were produced as fusion proteins with affinity handles, the purification process was also greatly facilitated. A requirement for this concept to succeed is that the disulfides of the produced proteins are allowed to break and reform. Therefore a redox buffer was added to the cultivation media. In contrast to earlier attempts to co-express "helping proteins" this system has shown to result in quantitative yields of correctly folded IGF-I.

PROTEINS TO FACILITATE RECOVERY

Recombinant DNA techniques allow fusion of genes or fragments thereof to alter the properties of recombinant proteins. The most widespread use of this strategy is to enable affinity purification of the product. It is possible to fuse the protein of interest to different extensions, proteins or peptides, in order to facilitate the purification procedure. Furthermore, by using DNA-technology it is possible to affect both the amount and the localization of the product. Effective systems for secretion of the protein to the medium or to the periplasm, give us the opportunity of achieving an initial purification step. The expression level of recombinant proteins has in several cases been increased by fusion and the number of other useful applications is constantly increasing. Gene fusions offer means to improve the proteolytic stability of sensitive recombinant proteins. Ubiquitin, well known for its resistance to proteases, can stabilize proteins in *E. coli*. Fusion to the IgG-binding tail ZZ has allowed recovery of full-length proinsulin and IGF-I. Also, a region consisting of several hydrophilic amino acids, fused to the C-terminus has been shown to stabilize a labile form of the phage repressor protein P22 Arc in the *E. coli* cytoplasm. To enable specific recovery of full length protein, a dual affinity system also has been developed by

TABLE 7.1: EXAMPLES ON AVAILABLE PROTEINS OR PEPTIDES FOR SELECTIVE PURIFICATION

Fusion partner	Ligand	Interaction	Elution conditions	Reference
Protein A (SPA)	hIgG/Fc	protein-protein	low pH	(Stahl and Nygren, 1997)
Z	hIgG/Fc	protein-protein	low pH	(Stahl and Nygren, 1997)
ProteinG (SPG)	hIgG/Fc	protein-protein	low pH	(Stahl and Nygren, 1997)
ABP	HSA	protein-protein	low pH	(Stahl and Nygren, 1997)
Glutathione-S-transferase (GST)	glutathione	protein-substrate	reduced glutathione	(Smith and Johnson, 1988)
Maltose-binding protein (MBP)	amylose	protein-carbohydrate	maltose	(Di Guan *et al*,. 1988)
FLAG peptide	mAB	peptide-protein	low pH/FLAG pept.	(Hopp *et al*., 1988)
Biotinylated protein	Monoaviding	biotin-avidin	Biotin	(Wilchek and Bayer, 1990)
polyhistidine	Me2+	peptide-IMAC	low pH/imidazole	(Hochuli, 1990, Hochuli *et al*., 1988)
polyarginine	anion	peptide-ionexchanger	salt	(Sassenfeld and Brewer, 1984)
polytryptophane	–	2-phase system	–	(Kohler *et al*., 1991)

Hammarberg *et al*. The protein of interest is fused with two different affinity tags that separately can be used in affinity chromatography. This system has also proven to further stabilize the protein compared to single affinity fusion. By what mechanism a protein fusion inhibnits proteolysis remains unclear in many cases. Suggested mechanisms include (i) protected N-or C-terminus, (ii) sterical hindrance from proteases and (iii) improved folding.

A variety of different gene fusion systems in order to use affinity chromatography have been developed since the first example was presented. Different proteins show very divergent characteristics, regarding stability, solubility, size and secretability. Therefore, proteins need individual treatment and no ideal fusion system exists that is applicable on all proteins.

Fusion protein purification systems can be based on several kinds of interaction. Examples of interactions are protein-protein interactions, enzyme-substrate interactions, protein-carbohydrate interactions and protein-metal interactions. Some of the most frequently used fusion protein systems for affinity purification are glutathion-S-transferase (GST), protein A (SPA), maltose binding protein (MBP) and thioredoxin (TR). There are also a number of short peptides used for selective purification, such as poly-histidine, poly-arginine, and tryptophan tags. When choosing a system for affinity purification there are several factors to consider. The choice of system will influence the conditions for purification and elution. Also the size and stability of the fusion partner, the stability and regeneration possibility of the affinity matrix and the cost of the affinity matrix has to be considered. Recently developed fusion partner, and more attention is being paid to mild elution conditions. Important is also whether the fusion partner has to be removed or not. In many cases this is crucial and there are fusion partner has to be removed or not. In many cases this is crucial and there are several cleavage methods, chemical and enzymatic, available.

8

Cutinase-A Lipolytic Enzyme

Cutinases are enzymes of 22–26 KDa which are able to hydrolyse ester bonds in the cutin polymer. Cutinases are produced by several phytopathogenic fungi and pollen. For many years, cutinases have been thought to play a crucial role in the fungal invasion of host plants: the hydrolysis of cutin, an insoluble monocarboxylic acid polyester representing the main component of the plant cuticle, has been thought to enable the penetration of the fungus in the plant. However, interpretations of recent data on fungal cutinase activity and pathogenicity are contradictory, and range from cutinase having little or no influence at all on pathogenicity, to it enhancing adhesion of fungal spores to the plant surface. In addition to cutin polymer degradation, cutinases are also able to hydrolyse a large variety of synthetic esters and show activity on short and long chains of emulsified triacylglycerols as efficiently as pancreatic lipases. Contrary to lipases, the activity of which is greatly enhanced in the presence of a lipid/water interface, cutinases do not display, or display little, interfacial activation, being active on both soluble and emulsified triglycerides.

Cutinase from the fungus *Fusarium solani* is a 22 KDa enzyme which has been extensively investigated from both biochemical and structural standpoints. Since its 3D structure was solved at

1.6 Å (R-factor 15.4%), considerable efforts have been devoted to the elucidation of the catalytic mechanism and of the involvement of specific residues in substrate binding and catalysis, as well as to the rational design of mutants endowed with a better catalytic performance.

TABLE 8.1: (A) SPECIFIC ACTIVITY OF CUTINASE, HPL AND RML AS A FUNCTION OF THE CHAIN LENGTH OF THE SUBSTRATE. (B) SPECIFIC ACTIVITY OF CUTINASE AND OF RML ON TRIBUTYRIN BELOW AND BEYOND THE CMC

a

U/mg	Tributyrin (C4)	Triolein (C18:1)
Cutinase	3 200 (40%)	670 (14%)
hpL	8 000 (100%)	4 600 (96%)
RmL	8 000 (100%)	4 800 (100%)

b

U/mg (tributyrin)	Below CMC	Beyond CMC	Ratio
Cutinase	1 280	3 200	25
RmL	53	8 000	150

During the last six years, the structures of 35 cutinase variants, including covalently inhibited complexes, as well as rationally designed point mutants, have been determined, and the resolution of native cutinase was extended to 1.15 Å (R-factor 16.7%). During all these investigations, striking evidences on the stability of cutinase were obtained, thus making this enzyme an excellent candidate to collect data on high energy X-ray sources. Accordingly, monochromatic and Laue data were collected on different synchrotron sources.

In this review we describe the most relevant information arising from this considerable body of structural information.

THREE-DIMENSIONAL STRUCTURE OF NATIVE CUTINASE

The Overall Structure

Crystals of recombinant *Fusarium solani* native cutinase crystallize in space group $P2_1$. No density could be observed for 16 amino acids at the N terminus and two at the C terminus. Residues 1 to 15 are part of a propeptide which is not present in the natural mature enzyme, which could explain their high extent of disorder. Cutinase is a compact one domain molecule (45 * 30 * 30 $Å^3$) consisting of 197 residues. It has a central β-sheet consisting of 5 parallel β-strands covered by 2 α-helices on each of the two sides of the sheet. The active site — consisting of the catalytic triad Ser 120, Asp 175 and His188 — is reminiscent of that found in serine proteases, but with an opposite handedness. Unlike RmL, hpL and GcL, the catalytic serine is not buried under an amphipathic loop but is accessible to the solvent: it is located at one extremity of the protein ellipsoid at the bottom of a crevice delimited by the loops 80–87 and 180–188. These loops, which bear hydrophobic amino acids, constitute the lipid binding site (LBS) of cutinase. The extended active site crevice is partly covered by two thin bridges formed by the side chains of residues Leu 81 and Val 184, and Leu 182 and Asn 84, respectively. The presence of a catalytic site directly exposed to the solvent may provide a structural explanation for the absence of interfacial activation. The side chains of the hydrophobic amino acids belonging to the LBS display B-factor values higher than the average. This may be an indication that these side chains are mobile and could move upon interface binding. Based on the 3D structure of the native, uncomplexed enzyme it was postulated that the binding of cutinase to the interface would not require major main chain rearrangements but only the re-orientation of a few lipolytic side chain that would play the role of a mini-lid. The accessibility of the catalytic serine would provide a structural explanation for the esterase activity on short triglycerides (tributyrin, C4) at substrate concentration below the critical micellar concentration (CMC).

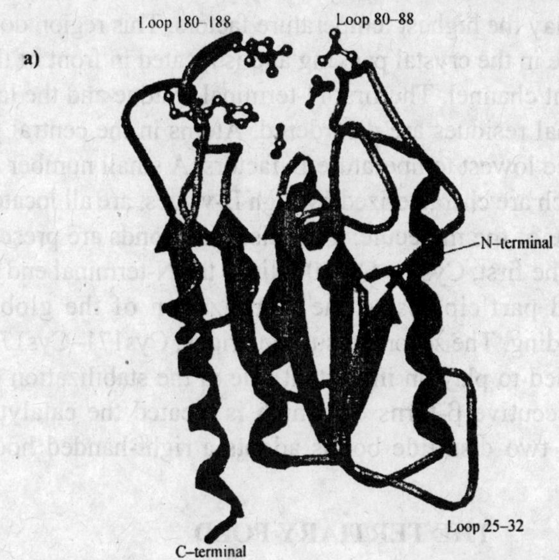

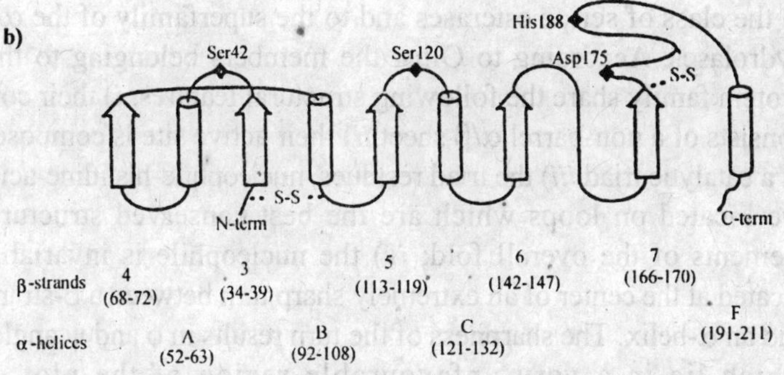

Figure 8.1: ***(a) Ribbon representation of cutinase. α-helices (red), β-strands (blue) and coils (yellow) are shown. The residues of the catalytic triad (Ser 120, His 188, Asp 175), as well as the residues constituting the two thin bridges over the catalytic crevice (Leu 81, Val 184, Leu 182 and Asn 84) are represented in balls and sticks (atom type colouring). The loops delimiting the catalytic crevice (residues 80–88 and 180–188) as well as the loop 25–32 are highlighted. (b) Topology representation of cutinase. Disulfide bridges are represented by dashed lines, strands by arrows and helices by cylinders. The catalytic residues are represented by diamonds***

Besides the two loops delimiting the catalytic crevice, the loop 25–32, which is located far away and opposite to the LBS, was found to display the highest temperature factors. This region does not participate in the crystal packing and is located in front of the largest solvent channel. The first N-terminal residue and the last two C-terminal residues are disordered. Atoms in the central β-sheet show the lowest temperature B-factors. A small number of residues, which are characterized by high B-values, are all located on the surface of the molecule. Two disulfide bonds are present in cutinase. The first, Cys31–Cys109, links the N-terminal end to a β–turn and participates to the stabilization of the global molecular folding. The second disulfide bridge, Cys171–Cys178, can be assumed to play an important role in the stabilization of the two consecutive β-turns on which is located the catalytic Asp175. The two disulfide bonds adopts a right-handed hook conformation.

THE TERTIARY FOLD

Cutinase is the smallest lipase/esterase structure solved so far. Like all the lipolytic enzymes investigated so far, cutinase belongs to the class of serine esterases and to the superfamily of the α/β hydrolases. According to Ollis, the members belonging to this protein family share the following structural features: *i*) their core consists of a non-barrel α/β sheet *ii*) their active site is composed of a catalytic triad *iii*) the triad residues, nucleophile-histidine-acid, are located on loops which are the best conserved structural elements of the overall fold; *iv*) the nucleophile is invariably located at the center of an extremely sharp turn between a β-strand and an α-helix. The sharpness of the turn results in ϕ and ψ angles which lie in a very unfavourable region of the plot of Ramachandran (ε conformation). The occurrence of an α/β hydrolase fold and of a catalytic machinery composed by a nucleophile, an acid and an histidine, seems to be a common feature to all the esterases described so far, with the only exception represented by *Streptomyces scabiaes* esterase.

Cutinase, which is the smallest member of the β/α hydrolase superfamily, is an α/β protein with an hydrophobic core comprising a slightly twisted five-parallel-stranded β-sheet, surrounded by four α-helices. The catalytic serine (Ser 120) is located at the top of a very sharp γ–like turn between the strand β5 and the helix αC. The sharpness of this turn, which is a type II' turn, results in the energetically unfavourable ε–conformation of Ser 129 (ϕ= 59°, ψ = – 122°) typical of the nucleophilic residue of all the members of the α/β hydrolase fold.

TABLE 8.2: STRUCTURES OF THE A/B HYDROLASE FOLD FAMILY MEMBERS

Enzyme source-complexed with	Abbreviation	Res. (Å)	Reference	PDB Code
Lipase				
human pancreatic	hpL	2.3	Winkler *et al.*, 1990	n.d.*
-porc colipase-C1 phosphonate	hpL-COL-C11	2.45	Egloff *et al.*, 1995	1LPB$
horse pancreatic	hopL	2.3	bourne *et al.*, 1994	1HPL*
guinea pig (N-term)	gpL		Martinez *et al.*, 1996	*
Rhizomucor miehei	RmL	1.9	Brady *et al.*, 1990	3TGL*
-n-hexylphosphonate ethyl ester		1.9	Brzozowski *et al.*, 1991	5TGL$
Penicillium camembertii	PcL	2.0	Detewenda *et al.*,1994a-c	1TIA
Humicola lanuginosa	HIL	1.84	Derewenda *et al.*, 1994a-b-c	1TIB
Rhizopus delemar	RdL	2.6	Derewenda *et al.*, 1994a-b-c	1TIC*
Geotrichyum candidum	GcL	1.8	Schrag & Cygler, 1993	1THG*
Candida rugosa	CrL	2.06	Grochulsky *et al.*, 1993	1CrL*
(1S)-menthyl hexyl phosphonate	(1S)-CrL-MPC	2.2	Grochulsky *et al.*, 1994a	1LPS$
Pseudomnas mendocina	Bott *et al.*,1993	n.d.		
Pseudomonas glumae	PGL	3.0	Noble *et al.*, 1994	1TAH*
Candida antarctica B	CaBL	1.55	Uppenberg *et al.*, 1994	1TCA$
Chromobacterium viscosum		3.0	Lang *et al.*, 1994 & 1996	n.d.
Cutinase				
Fusarium solani pisi	FsC	1.6	Martinez *et at.*, 1992	1CUS*
-diethylphosphate	FsC-E600	1.9	Martinez *et al.*, 1994	2CUT$
Esterase				
Torpedo californica acetylcholinesterase	AChE	2.8	Sussmn *et al.*, 1991	1ACE*
-tacrine	AChE-THA	2.8	Harel *et al.*, 1993	1ACJ$
Pseudomonase fluorescens esterase	PfE		Se Won Suh, unpublished	n.d.*
Vibrio harveyi thioesterase	TE	2.1	Lawson *et al.*, 1994	1THT*
Bromoperoxydase				

Streptomyces aureofaciens A2		2.05	Hecht *et al.*, 1994	
Serine carboxypeptidase				
Triticum vulgaris II	CPD-WII	2.2	Liao & Remington, 1992	3SC2*
-L-benzylsuccinate 1WHT$	CPD-WII-BZS	2.0	Bullock *et al.*, 1994	
Saccharomyces cerevisiae	CPD-Y	2.8	Endrizzi *et al.*, 1994	1YSC*
Dienelactone hydrolase				
Pseudomonas sp. B13	DHL	2.8	Pathak *et al.*, 1988	n.d.
Haloalkane dehalogenase				
Xanthobacter autotrophicus	HAL	1.9	Verschyueren *et al.*, 1993a	2HAD*
-2-chloroethane, pH = 5.0, 22° C	HAL-DCE	2.0	Verschueren *et al.*, 1993b	2DHD$
Protective protein				
Human	HPP	2.2	Rudenko *et al.*, 1995	

Are reported all the uncomplexed forms of the α/β hydrolase fold members and one of their complexed forms. *, $: atomic co-ordinates used in the structural comparison of the open and closed forms, respectively. n.d.: atomic co-ordinates are not deposited with the Brookhaven Protein Data Bank.

SECONDARY STRUCTURE

The secondary structure elements based on the definition of Kabsch & Sander (1983) have been numbered to emphasize the similarity in the secondary structures of the α/β hydrolase fold family. The five β-strands of the central β-sheet represent 15% of the total residues. 23 out of the 25 residues of the β-sheet, are hydrophobic and contribute to maintain a central water-excluded environment in the core of the protein. 6 out of the 12 residues located on the concave side of the β-sheet are aromatic or long side chain amino acids. On the convex side, there are 13 residues with shorter side chains. In the central β-sheet, the hydrogen bond pattern typical of parallel strands is observed, all the 17 hydrogen bonds occur between backbone atoms and fall within the expected arrange. There is no internal β-turn in the cutinase model.

The four crossover connections between the parallel strands are either of the common right-handed β/.α/β or β/ loop/βtype. In the proximity of the active site region, 57 residues form five α/β loops. The loop connecting the strand β-7 and the helix α F (residues 171–190) contains the catalytic Asp 175 and His 188 residues. On the opposite side of the active site, 31 residues form four α/β loops.

The four helices of the α/β core represent 31% of the total residues. Their length ranges from 11 to 21 residues. The four helices have the same orientation and are located on the surface of the molecule on each side of the central β-sheet. Helices α A and αF pack against the concave side of the β-sheet. They are parallel to strand β-3 and β-7, respectively. Helices α B and α C pack against the convex side of the β-sheet. The packing of the four α-helices against the β-sheet is mediated by a network of hydrophobic interactions involving 14 aromatic residues. No hydrogen bonds between either the β-sheet and the α-helices or two α-helices are observed.

Helices αA (11 residues), αB (17 residues), and αC (12 residues) are classical right-handed 3.6_{13} α-helices. The 3.6_{13} helix αF (21 residues) presents a pronounced kink with a distorted α-helix hydrogen bond pattern, caused by the occurrence of an additional residue. As a result, this helix accepts three proline residues. Pro 193 is in the turn preceeding the N-terminal part of the kink, while Pro 198 and Pro200 are distributed in the two first turns following the C-terminal of the kink. As a result, the α-helix hydrogen bond pattern is not classical, with the Asp 194 backbone oxygen atom of the kink being hydrogen bonded with the (n+5) th residue amide group. The four residues (194–197) of the kink resemble to a type III turn except for the turn hydrogen bond. As Asp 194 O is involved in the α-helix hydrogen bonding, its hydrogen bond with Gly 197 NH is in an unvafourable conformation and very weak (3.42 Å). The backbone oxygens of Ala 195 and Arg 196 do not participate in the α-helix hydrogen bonding pattern.

Among the three one-turn helices external to the α/β core, the helices αA (4 residues) and αC (6 residues) are 3.6_{13} α-helices, while the helix αB (5 residues) is the only 3_{10} helix of the molecule. The helix αA, located on the surface of the molecule, belongs to the N-terminal loop proceeding the first strand (β-3) of the central β-sheet. The helix αB connects the strand β-4 and the helix αB. The helix αC, located in the loop region opposite side to the active site, connects the helix α C and the strand β-7.

TABLE 8.3 : IONIC INTERACTIONS AND SIDE CHAIN TO SIDE CHAIN HYDROGEN BONDS IN CUTINASE

Residue 1		Distance (Å)	Residue 2	
Arg20	NH2	2.80	Oδ1	Asp22
	NH2	2.77	Oε1	Glu60
	NH1	2.86	Oε1	
Arg78	NH1	2.88	Oδ1	Asp83
Arg88	NH2	2.70	Oδ1	Asp132
	NH1	2.88	Oδ2	
	NH	12.95	Oε2	Glu131
Arg20	Nε	3.13	Oδ2	Asp22
Asp21	Oδ2	3.17	Nε2	Asn47
	Oδ2	2.92	Nε2	Gln71
Glu44	Oε2	2.72	Oγ1	Thr80
Arg78	Nε	2.92	Oγ1	Thr90
Asn106	Oδ1	2.74	NZ	Lys140
Thr113	Oγ1	2.87	NHI	Arg211
Thr144	Oγ1	3.18	OH	Tyr162
Asn155	Nδ2	2.99	OH	Tyr149
*Ser42	Oγ	3.12	Nε2	Gnl121
*Gln121	Oδ1	2.86	Oγ1	Thr90
*His188	Nε2	2.76	Oγ	Ser120
*His188	Nδ1	2.86	Oδ1	Asp175
*His188	Nδ1	3.14	Oδ2	

The upper and lower part of the table list the salt-bridges and the side chain hdrogen bonds, respectively. * indicates active site residus, including the three catalytic residues and the residues involved in the stabilization of the substrate oxyanion.

HYDROGEN BONDS AND SALT BRIDGES

In Table are listed the side chain to side chain hydrogen bonds and salt-bridge interactions between polar and charged residues.

All these interactions are found on the surface of the protein, except the hydrogen bond cluster of the active cavity. The 7 salt bridges form three clusters of ionic interactions, namely Arg20 with Glu60, Arg78 with Asp83, and Arg88 with Asp 132 and Glu131. Contrary to the RmL and hpL, cutinase has no internal charged residues. There are 183 main chain to main chain hydrogen bonds. None of the 253 solvent molecules present in the 1.15 Å model establish contact with the central β-sheet. Most of them are located on the protein surface near α-helices. The active site cavity contains two water molecules. There are five internal water molecules stabilized by hydrogen bonds with carboxyl and amide main chain groups of hydrophobic residues.

THE CATALYTIC TRIAD

The active site of cutinase is formed by the catalytic triad (Ser 120, Asp 175 and His 188), which is reminiscent of that found in serine proteases. However, in the α/β hydrolase fold enzymes, the twist of the sheet imposes a handedness on the catalytic triad which is approximately a mirror-image of that observed in the serine proteases. The amino acid sequence encompassing Ser 120 (Gly-Tyr-Ser-Gln-Gly) matches the consensus sequence Gly-His (Tyr)-Ser-X-Gly commonly present in lipase sequences. The catalytic residues are in the same sequence order observed in all the members of the α/β hydrolase fold family. These residues are located in turns that project their side chain away from the polypeptide backbone into accessible positions. Cutinase Ser 120 is located in a tight turn, classified as type II', between the C-terminal end of the strand β-5 and the helix α C. The sharpness of the turn forces the side chain od Ser 120 to point into the active site cavity. Asp 175 and His 188 are very close in sequence and are located on the same loop between the strand β-7 and the helix α F. His 188 is located at position 4 of a type I β-turn. As usually observed in this type of reverse turns, the residue proceeding the histidine is a proline. Asp 175 is located at position 1 of a type I turn. The carboxylate of Asp 175 is stabilized by means of two hydrogen bonds to the backbone nitrogens of the residues Cys 178 and Val 177. All hydrogen bonds identified in the active site are characteristic of the serine proteases: Ser 12 Oγ is bonded to His 188 Nε2 (2.76 Å) while Nδ1 of the imidazole ring is hydrogen bonded to Asp 175 Oδ2 (3.14 Å) and Oδ1 (2.86 Å).

STRUCTURAL COMPARISON BETWEEN CUTINASE AND THE OTHER α/β HYDROLASE FOLD MEMBERS

Even if the members of the α/β hydrolases fold family vary widely in their sizes, amino acid sequences, substrates and physical properties, they present an appreciable structural similarity. As already mentioned, among the common features defining the α/β hydrolase fold, the key points are the central β-sheet, the presence of a catalytic triad and the location of the nucleophile active site residue in a β-hairpin. Catalytic triad and the location of the nucleophile active site residue in a β-hairpin. Despite these common features, the α/β hydrolase fold members present significant differences. The differences include variations of *i*) the number of strands in the central β-sheet, *ii*) the nucleophilic and acidic residue type of the catalytic triad, and *iii*) the number of external central core motifs.

The secondary structure labels have been chosen to emphasize the similarity in the secondary structures of the different members of the α/β hydrolase fold family. The structural elements of the central core have been used to define eleven topological groups. The groups are classified based on the number of α/β hydrolase motifs; the group I is the smallest member of the family with five strands and four helices while the group XI is the largest one with eleven stands and six helices. Five groups (VII to XI) contain the central eight stranded motif defined by Ollis *et al*, (1992). HIL, PcL, RdL, and RmL which belong to a same enzymatic family are distributed into the topolocical groups II and IV. The consensus topology of the central β-sheet is: –1x, +2x, +1x, +1x, + 1x. The groups V, VI, VII, VIII, IX, and XI show a complete agreement with this consensus topology, while the groups I, II, III, IV, and V present small differences in the external part of the central core. The best representant of the consensus α/β hydrolase fold topology if the group VII which does not present extra motifs.

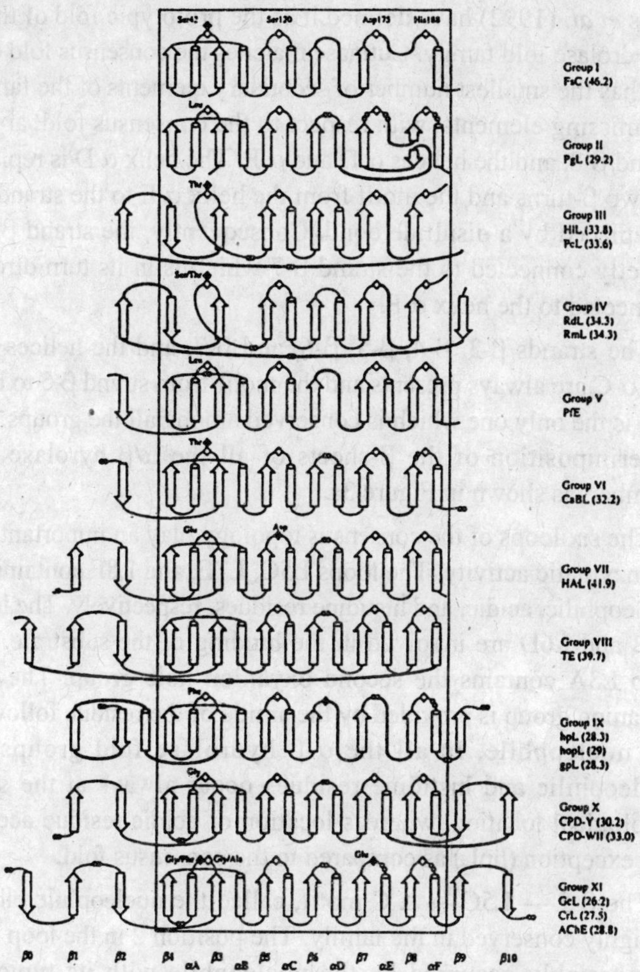

Figure 8.2 : ***The eleven topologic groups of the α/β bydrolase fold proteins. The percentages of the residues belonging to the consensus α/β hydrolase fold (out of the total residues) are shown in brackets. The α6/β6 consensus motif is contoure. Consensus and extra loops are shown in thin and in thick, respectively. Extra β-strands or α-helices are represented as two small circles. Catalytic residues are represented by empty diamonds, except for the acidic member of groups IX and XI which are represented by a solid diamond. The second oxyanion group (main chain) is represented by an empty diamond. A solid diamond is added when the residue serves also as the third component of the oxyanion hole (side chain)***

Ollis *et al.* (1992) have defined it as the prototypic fold of the α/β hydrolase fold family. Cutinase matches the consensus fold even if it has the smallest number of secondary elements of the family; the missing elements, with respect to the consensus fold, are the strand β-8, and the helices α D and α E. The helix α D is replaced by two β-turns and the motif from the helix α E to the strand β-8 is replaced by a disulfide bond. Consequently, the strand β-6 is directly connected to the strand β-7 which is in its turn directly connected to the helix α F.

The strands β-3, β-4, β-5, β-6, and β-7, and the helices α B and α C are always presents and the motif from strand β-5 to helix α C is the only one which is conserved among all the groups. The superimposition of the β-sheets of all the α/β hyrolase fold members is shown in Figure 3a.

The six loops of the consensus topology play an important role in enzymatic activity. The loops L5C, L7E, and L8F contains the nucleophilic, acidic, and histidine residues, respectively. The loops L4B and L6D are involved in the binding of the substrate. The loop L3A contains the second oxyanion hole group. The first oxyanion group is provided by the amide of the residue following the nucleophile. In all the α/β hydrolase fold groups the nucleophilic and histidine residues occur always at the same topological location, whereas location of acidic residue accepts one exception (hpL) as compared to the consensus fold.

The β-5 — L5C — α C motif, called the nucleophile elbow, is highly conserved in the family. The position 2 in the loop L5C is invariably occupied by the nucleophile with its typical ε conformation. In all α/β hydrolase fold proteins, the nucleophilic residue is a serine except in the case of HAL and of dienelactone hydrolase in which it is replaced by an aspartic and a cysteine, respectively.

The acidic residue is located in the loop L7E, except for the group IX in which it has a different topological origin, being located in the loop L6D. In all groups, the acidic residue is an aspartic, with the only exception being provided by the group XI in which it is replaced by a glutamic. The best conserved catalytic residue is the histidine which is conserved in all the members and which is invariably located in the loop L8F.

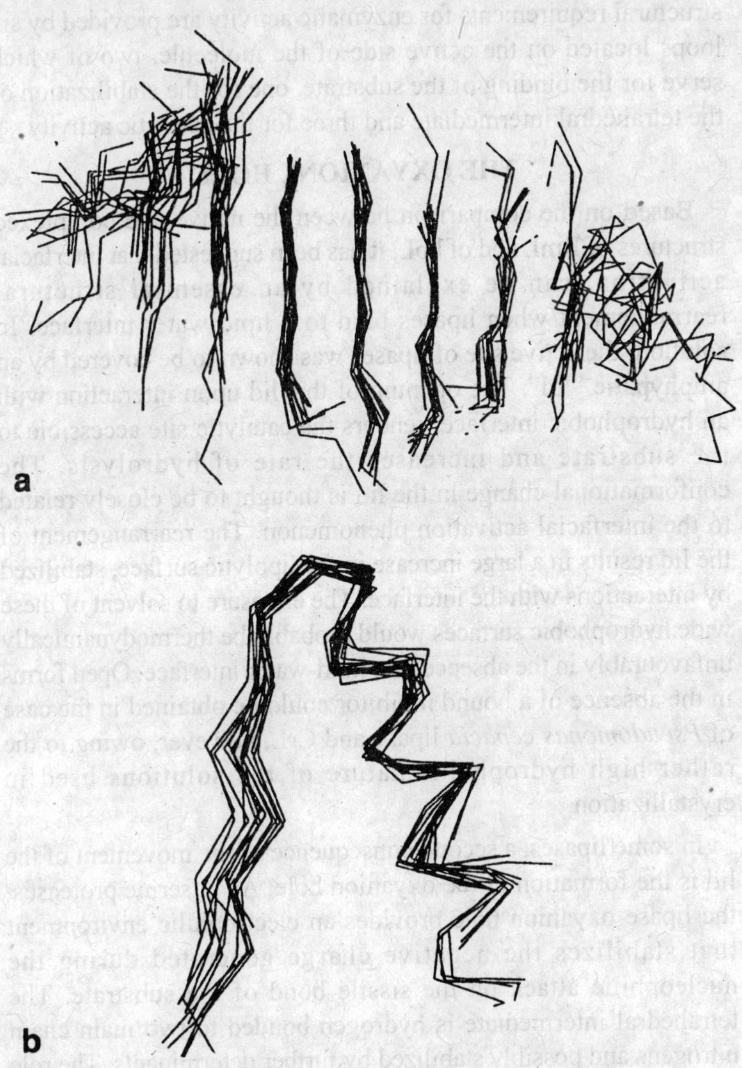

Figure 8.3 : *Superimposition (Cα tracing) of (a) the β-sheet and of (b) the nucleophile elbow of all the α/β hydrolase fold members*

Cutinase is presently the smallest α/β hydrolase member, with only five strands in the central β-sheet surrounded by four helices. The "minimal topology" of cutinase suggests that the essential structural requirements for enzymatic activity are provided by six loops located on the active side of the molecule, two of which serve for the binding of the substrate, one for the stabilization of the tetrahedral intermediate and three for the catalytic activity.

THE OXYANIONE HOLE

Based on the comparison between the native and complexed structures of RmL and of hpL, it has been suggested that interfacial activation, can be explained by an essential structural rearrangement when lipases bind to a lipid water interface. In solution, the active site of lipases was shown to be covered by an amphypatic "lid". The opening of this lid upon interaction with an hydrophobic interface, renders the catalytic site accessible to the substrate and increases the rate of hydrolysis. The conformational change in the lid is thought to be closely related to the interfacial activation phenomenon. The rearrangement of the lid results in a large increase in the lipolytic surface, stabilized by interactions with the interface. The exposure to solvent of these wide hydrophobic surfaces would probably be thermodynamically unfavourably in the absence of a lipid-water interface. Open forms in the absence of a bound inhibitor could be obtained in the case of *Pseudomonas cepacia* lipase and CrL, however, owing to the rather high hydrophobic nature of the solutions used in crystallization.

In some lipases, a second consequence of the movement of the lid is the formation of the oxyanion hole. As in serine proteases, the lipase oxyanion hole provides an electrophilic environment that stabilizes the negative charge generated during the nucleophilic attack on the sissile bond of the substrate. The tetrahedral intermediate is hydrogen bonded to two main chain nitrogens and possibly stabilized by further determinants. The role played by the amide nitrogen of the active site serine in the serine proteases is played by the amide nitrogen of the residue following the nucleophile in the α/β hydrolase superfamily. The relative position of the oxyanion hole residues is highly conserved in the α/β hydrolase superfamily: whereas the histidine and the acidic member of the catalytic triad are located on one side of the nucleophile (ser), the residues forming the oxyanion hole lie on the other side.

Phosphorylating agents, which block the serine protease reaction at the acylation step via the formation of a tetrahedral intermediate, provide a powerful tool to study one of the activated states of lipases/esterases and to characterise their oxyanion hole. Covalent complexes with organophosphate and organophosphonate inhibitors of four lipases and of cutinase were obtained and their structures were solved. These structures allowed to identify a relation between the position of the lid and interfacial activation. In most cases, the optimal oxyanion hole geometry is achieved only upon the conformational change which takes place after adsorption to the interface.

THE CUT-E600 COMPLEX

The comparison between the structures of native enzyme and of a covalently inhibited complex with *n* diethyl-*p*-nitrophenyl phosphate (E600) (1.9 Å) revealed that cutinase has a preformed oxyanion hole. The overall structure of the cut-E600 complex (crystallizing in a different space group) is very similar to that of the native enzyme. The only significant differences concept the loop 25–32 which is located opposite to the active site (Figure 4a) and which displays the highest B-factor values even in the native structure. The Ramachandran plots of native and complexed cutinase are almost identical indicating that no conformational changes occur upon inhibitor binding, contrary to what has been observed in the case of RmL and hpL. The E600 accommodation in the active site crevice takes place with only minor changes in the backbone and side chain positions of the loops 80–88 and 180–188 delimiting the active site crevice. The active site contains a water molecule which is bound with three main chain nitrogens (Leu 176 N, Ala 185 N and Ile 183 N) and to the catalytic Asp 175 Oδ1 atom, thus maintaining this residue

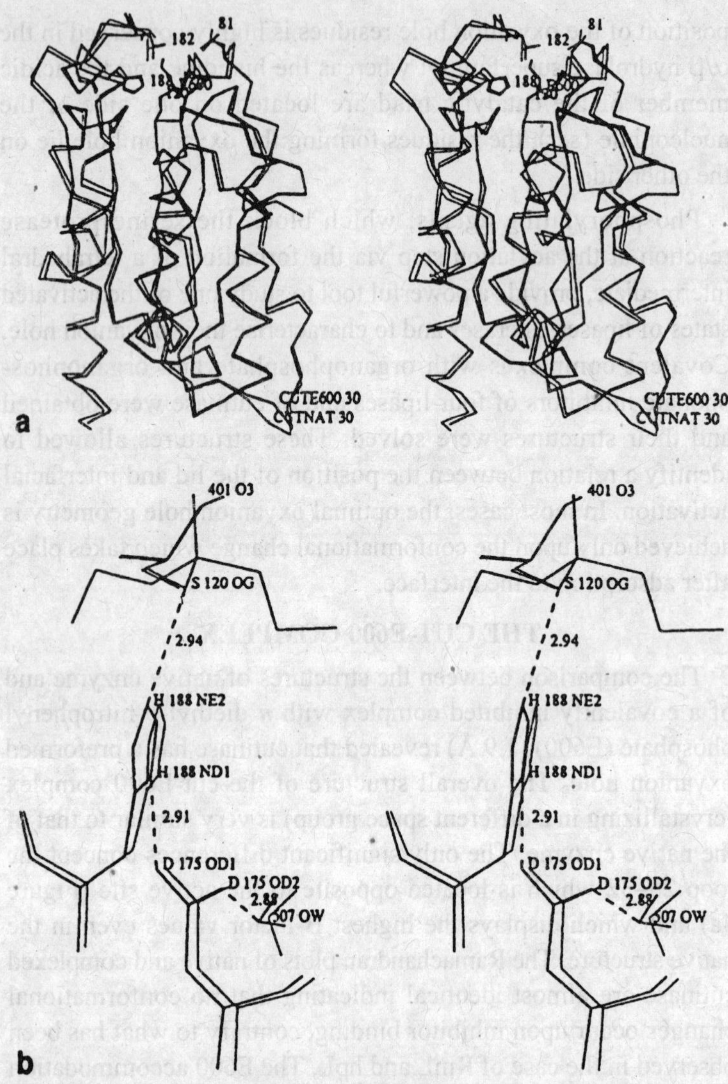

Figure 8.4: **(a)** ***Stereoview of the Cα tracing of cut-E600 (thick) superimposed on that of native cutinase (thin). The catalytic triad and the inhibitor are shown in balls and sticks. The loop 25–32 is highlighted.*** **(b)** ***Stereoview of the catalytic triad of cut-E600 (thick) superimposed on that of native cutinase (thin). The inhibitor, as well as the triad bound water molecule are shown.***

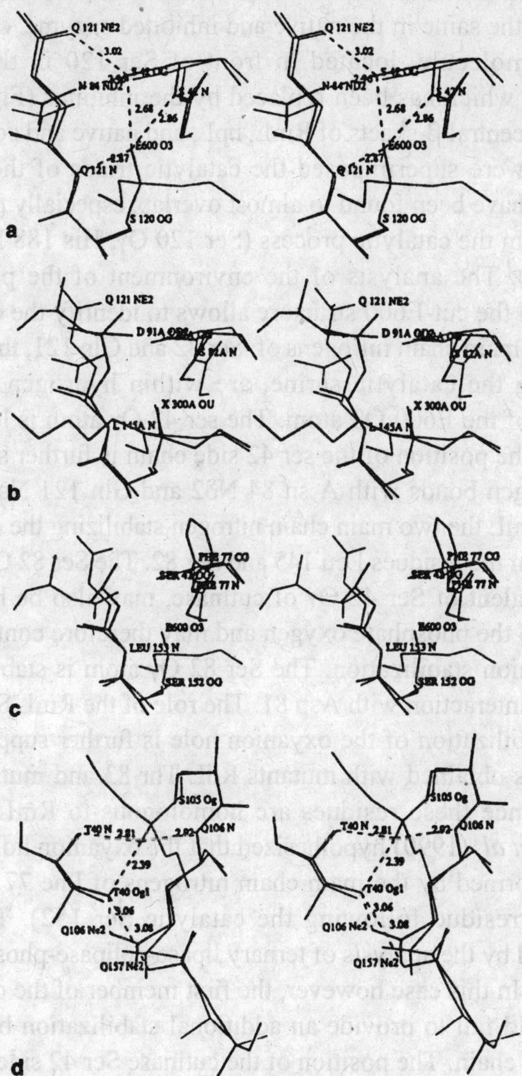

Figure 8.5: (a) ***Stereoview of the oxyanion hole of cut-E600 (thick) superimposed on that of native cutinase (thin). The E600 inhibitor as well as the hydrogen bonds involved in the oxyanion hole stabilization are shown. (b, c, d). Stereoview of the oxyanion hole of the inhibited RmL (b) (thick), that of the inhibited hpL (c) (thick) and that of the CaBL (d) (thick) superimposed on that of cut-E600 (thin). The residues involved in the oxyanion hole formation are shown***

in the proper orientation. The structure of the catalytic triad is basically the same in the native and inhibited enzyme, except for a water molecule, located in front of Ser 120 in the native structure, which has been replaced by the inhibitor (Figure 4b). After the central β-sheets of RmL, hpL, and native and complexed cutinase were superimposed the catalytic triads of these three enzymes have been found to almost overlap, especially the atoms involved in the catalytic process (Ser 120 Oγ, His 188 Nε2, Asp 175 Oδ1). The analysis of the environment of the phosphate oxygen in the cut-E600 structure allows to identify the oxyanion hole. The main chain nitrogens of Ser 42 and Gln 121, the residue following the catalytic serine, are within hydrogen bonding distance of the E600 O3 atom. The ser 42 Oγ atom is located at 2.68 Å. The position of the ser 42 side chain is further stabilized by hydrogen bonds with A sn 84 Nδ2 and Gln 121 Nε2. In the case of RmL the two main chain nitrogen stabilizing the oxyanion come from the residues Leu 145 and Ser 82. The Ser 82 Oγ, which the equivalent of Ser 42 Oγ of cutinase, may also be hydrogen bonded to the phosphate oxygen and may therefore contribute to the oxyanion stabilization. The Ser 82 Oγ atom is stabilized by only one interaction with Asp 81. The role of the RmL Ser 82 Oγ in the stabilization of the oxyanion hole is further supported by the results obtained with mutants RdL Thr 83 and mutants PcL Ser 84 since these residues are homologous to RmL Ser 82. Winkler *et al.* (1990) hypothesized that the oxyanion hole in hpL may be formed by the main chain nitrogens of Phe 77 and Leu 153 (the residue following the catalytic Ser 152). This was confirmed by the analysis of ternary lipase-colipase-phospholipid complex. In this case however, the first member of the oxyanion hole would fail to provide an additional stabilization by means of its side chain. The position of the cutinase Ser 42 side chain is also conserved in CaBL with Thr 40 Oγ1.

The atoms involved in the oxyanion hole of cutinase do not undergo conformational changes upon inhibitor binding, contrary to what has been observed in the case of RmL and of hpL, thus indicating that in cutinase a preformed oxyanion hole is present.

THE OXYANION HOLE MUTANTS

Based on the structure of the cut-E600 complex, it has been suggested that the cutinase oxyanion hole may consist not only of two main chain nitrogen atoms (Ser 42 N and Gln 121N) but also of the Ser 42 Oγ side chain. Moreover, Asn 84 N δ2 and Gln 121 Nε2 maintain Ser 42 Oγ in the proper orientation for the hydrogen bonding of the oxyanion to occur. Point mutants that selectively disrupt each of these interactions were designed in order to assess the potential function of the Ser42 side chain as part of the cutinase oxyanion hole. To evaluate the extent of the tetrahedral intermediate stabilization provided by Ser 42 Oγ, Ser 42 has been mutated into alanine. Since the proper orientation of Ser 42 Oγ is directed by Asn 84, the mutants Asn 84 to Ala, Leu, Asp and Trp, respectively, were produced to investigate the importance of this indirect interaction in the stabilization of the oxyanion hole. Biochemical modifications of cutinase mutants have been evaluated with regard to their ability to hydrolyze *para*-nitrophenyl butyrate (pnp-butyrate) in solution, since cutinase is able to hydrolyze this soluble substrate. Any secondary effects originating from interfacial activation can therefore be dissociated from the primary effects on the integrity of the oxyanion hole.

TABLE 8.4: STEADY STATE KINETIC PARAMETERS OF NATIVE AND MUTANTS CUTINASE DURING THE HYDROLYSIS OF PNP-BUTYRATE

	$k_{cat}s^{-1}$	$k_{cat}/K_M s^{-1}.mM^{-1}$	% residual activity
WT	1800	2647	100.00
S42A	4	12	0.22
N84A	478	1062	26.5
N84L	55	190	3.00
N84W	2	6	0.11
N84D	2.5	19	0.16

The S42A mutation resulted in a drastic decrease in the activity (450 fold, see Table 8.4) without significantly perturbing the three dimensional structure and the active site geometry. Mutations N84W, N84L and N84A disrupt the hydrogen bond stabilization of Asn84 Nδ2 with Ser 42 Oγ and modify the steric environment in the active site pocket. The activity is reduced to 0.11 % in the Trp mutant and remains at 3% and 26.5% with Leu and Ala mutants, respectively. As the result of these three mutations with non polar residues, the catalytic triad and the oxyanion hole geometry remain unchanged (Figure 6 b, c, d). The introduction of a net negative charge into the oxyanion hole pocket resulting from the N84D mutation modifies the hydrogen bonding network in the catalytic triad and the oxyanion hole and the activity is reduced to 0.13%. Although Asn84Leu and Asn84Ala retain some esterase activity. Our kinetic experiments and structural analysis of these cutinase mutants provide new evidence supporting the idea that Ser42 Oγ is an extra component of the cutinase oxyanion hole, acting in combination with the two main chain nitrogen atoms of residues Ser 42 and Gln 121. However, the reason (s) why cutinase and some lipases need three hydrogen bonds to stabilize the oxyanion, and subtilising and other lipases need only two, still remains to be elucidated.

THE CUT-TC4 COMPLEX

Although studies based on the use of small organophosphate and organophosphonate inhibitors have widely contributed to the elucidation of the catalytic mechanism of lipases, these compounds bear little resemblance to a natural triglyceride substrate.

In this context, the resolution of RmL complexed to a C6-phosphonate (*n*-hexyl *o*-ethyl chloro phosphonate, EECP), of CrL complexed with long alkyl sulfonyl chains and of a human pancreatic lipase-colipase complex covalently inhibited by a long (C11) alkyl chain phosphonate represent important attempts to mimic more closely and more realistically tetrahedral intermediates.

Taking into account the marked preference of cutinase for short chain (C4) triglycerides, a four carbon phosphonate inhibitor (TC4) that mimics 1-2 trihexanoyl-3 pentanoyl glycerol has been

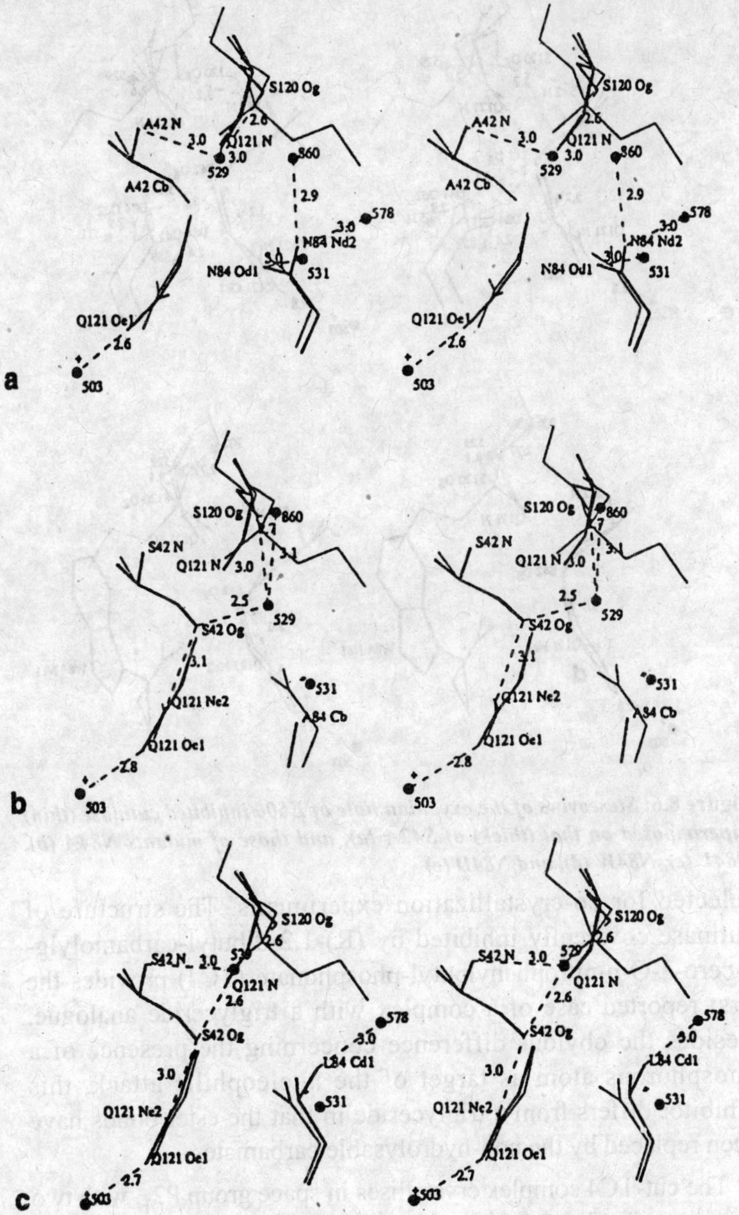

S120 Og
2.6
A42 N
3.0
Q121 N
3.0
860
529
A42 Cb
2.9
3.0
578
N84 Nd2
3.0
N84 Od1
531
Q121 Oe1
2.6
503

a

S120 Og
860
2.7
S42 N
3.1
Q121 N
3.0
2.5
529
S42 Og
3.1
531
Q121 Ne2
A84 Cb
Q121 Oe1
2.8
503

b

S120 Og
2.6
S42 N
3.0
529
Q121 N
2.6
578
S42 Og
3.0
L84 Cd1
3.0
531
Q121 Ne2
Q121 Oe1
2.7
503

c

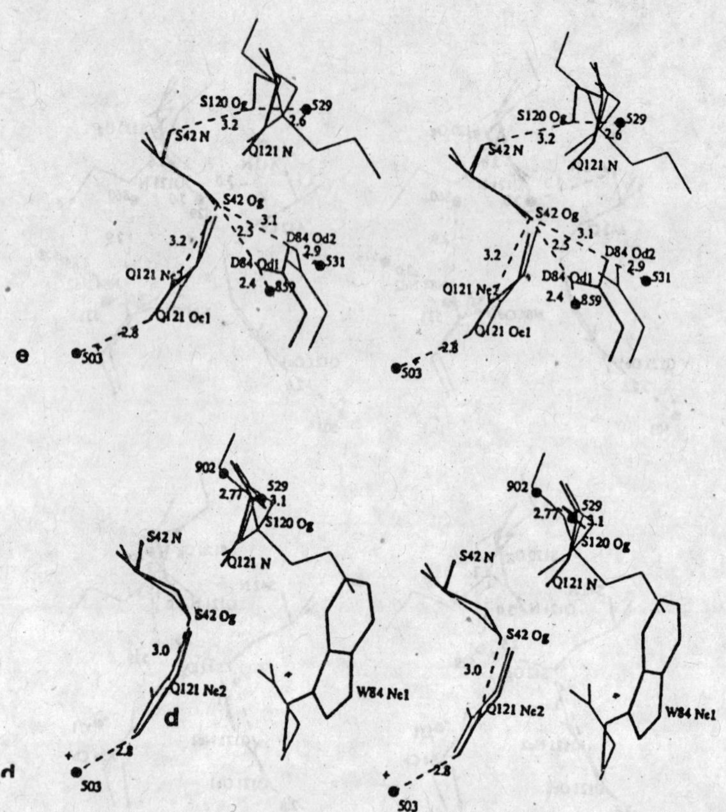

Figure 8.6: ***Stereoviws of the oxyanion hole of E600 inhibited cutinase (thin) superimposed on that (thick) of S42A (a), and those of mutants N84A (b), N84L (c), N84W (d), and N84D (e)***

selected for co-crystallization experiments. The structure of cutinase covalently inhibited by (R)-1,2-dibutyl-carbamolylglycero-3-O-*p*-nitrophenylbutyl-phosphonate (TC4) provides the first reported case of a complex with a triglyceride analogue. Besides the obvious difference concerning the presence of a phosphorous atom as target of the nucleophilic attack, this inhibitor differs from a triglyceride in that the ester bonds have been replaced by the non-hydrolysable carbamates.

The cut-TC4 complex crystallises in space group $P2_1$, with two protein molecules (Mol A and Mol B) in the asymmetric unit. The

TABLE 8.5: EXAMPLES OF OXYANION STABILIZATION COMPONENTS

Nucleophile-His-Asp (Glu) hydrolase with its nucleophilic residue			Main chain group 1st NH	2nd NH	3rd NH	Side chain group NH	OH
Serine Protease	Trypsin	(S195)	S195	G193			
	Chymotrypsin	(S195)	S195	G193			
	Proteinase A	(S195)	S195	G193			
	Subtilisin	(S221)	S221			N155 Nδ2	
	Carboxypeptidase II	(S146)	S147	G53			
Cysteine Protease	Papain	(C25)	C25			Q19 Nε2	
Esterase	*Streptomyces scabies esterase*	(S14)	S14	G66		N106 Nδ2	
	Acetylcholinesterase	(S200)	A201	G118	G119		
	Cholesterol esterase	(S209)	A210	G124	G123		
Lipase	*Candida rugosa* lipase	(S209)	A210	G124	G123		
	Human pancreatic lipase	(S152	L153	F77			
	Horse pancreatic lipase	(S152)	L153	F77			
	Pseudomonas glumae lipase	(S87	Q88	L17			

Table 8.5 Contd.

Table 8.5 Contd.

Nucleophile-His-Asp (Glu) hydrolase with its nucleophilic residue		Main chain group			Side chain group	
		1st NH	2nd NH	3rd NH	NH	OH
Mucor miehei lipase	(S144)	L145	S82			S82 Oγ
Penicillium camembertii lipase	(S145)	L146	S81			S81 Oγ
Humicola lanuginosa lipase	(S146)	L146	T83			S83 Oγ
Rhizopus delemar lipase	(S145)	L146	T83			T83 Oγ1
Candida antartica B lipase	(S107)	Q106	T40			T40 Oγ1
cutinase	(S120)	Q121	S42			S42 Oγ

overall structure of both molecules of the cut-TC4 complex is very similar to that of the native enzyme. As in the case of the cut-E600 complex, the central β-sheet is the best conserved part and the most important Cα deviations occur in the C-terminus and in the loop 25–32 opposite to the substrate binding site. As already observed with the cut-E600 complex, only minor differences could be observed between the lipid binding sites of both molecules and that of the native cutinase. The absence of large conformational changes in the substrate binding region upon binding to a non-hydrolysable analogue, represents to date a feature unique to cutinase, contrary to the large movements that have been reported for covalently inhibited complexes of other lipases.

The two protein molecules of the asymmetric unit are oriented head-to-head with their substrate binding sites facing and interacting each other. This structural arrangement allows the large hydrophobic surface, formed by the substrate binding region itself and by the inhibitor molecule, to be shielded from the solvent. The burying of hydrophobic surfaces in the crystal lattice by means of dimerization has been also observed in other lipase/esterase complexes as well as in an open form of CrL.

Only the S_P enantiomer of the inhibitor could be fitted into the electron density map of both sites. This observation suggests that only one enantiomer at phosphorous of the TC4 inhibitor had reacted, in agreement with the kinetic data showing that one of the stereoisomers at phosphorous reacts 60-fold faster than the other one. In both molecules, the overall shape of the inhibitor is that of a fork: the two dibutyl carbamoyl chains point toward the surface of the protein, while the butyl chain, linked to the phosphorus atom, is roughly perpendicular to the two dibutyl-carbamoyl chains. The butyl chain is embedded in a buried position at the bottom of the active site crevice, while the dibuyl-carbamoyl chains occupy a more exposed position. The while inhibitor molecule is almost entirely embedded in the active site crevice, with only C10, C11, C16 and C17 atoms protruding from the surface of the protein.

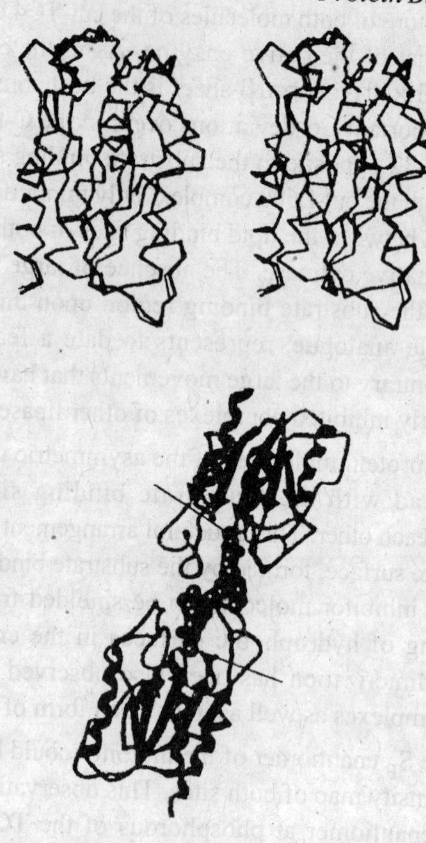

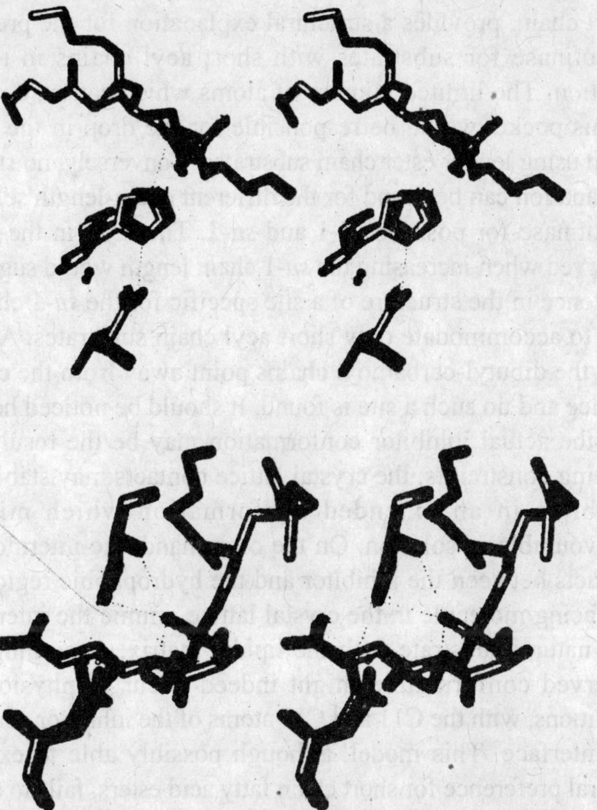

Figure 8.7: **(a)** ***Stereo view of the superimposition of the Ca tracing of Mol B (yellow) on the native structure (cyan). The TC4 B molecule is also shown (atom type colouring). (b) Secondary structure ribbon representation of the cut-TC4 complex, showing the two molecules in the asymmetric unit (Mol A and Mol B). The two inhibitor molecules are shown in CPK. (c) The lipid binding site region of Mol B. The solvent accessible surface (blue), the inhibitor (cpk), the loops 80–87 and 180–188 surrounding the catalytic crevice (yellow), the central β-sheet (blue) and the α-helices (red) are shown. (d) Stereo view of the catalytic triad of Mol B (atom type colouring) superimposed on that of the native enzyme (blue). The TC4 B (atom type colouring) and the E600 (magenta) inhibitors are displayed. The triad bound water molecule of the TC4 inhibited cutinase (red) and of the native enzyme (blue), together withy the hydrogen bond network are also shown. (e) Stereo view of the oxyanion hole of Mol B (atom type colouring) superimposed on that of the native enzyme (blue). The TC4 inhibitor (atom type colouring) as well as the hydrogen bonds involved in the oxyanion hole stabilization are also shown***

The presence of a rather small pocket which accomodates the butyl chain, provides a structural explanation for the preference of cutinase for substrates with short acyl chains in the *sn*-3 position. The limited number of atoms which can be embedded in this pocket would be responsible for the drop in the activity when using longer ester chain substrates. Conversely, no structural explanation can be found for the different chain-length selectivity of cutinase for position *sn*-1 and *sn*-2. The drop in the activity observed when increasing the *sn*-1 chain length would suggest the existence in the structure of a site specific for the *sn*-1 chain and able to accommodate only short acyl chain substrates. Actually, both the dibutyl-carbamoyl chains point away from the catalytic crevice and no such a site is found. It should be noticed however, that the actual inhibitor conformation may be the result of the packing constraints: the crystal lattice contacts may stabilise the inhibitor in an extended conformation which might be unfavourable in solution. On the other hand, the intermolecular contacts between the inhibitor and the hydrophobic region from the facing molecule in the crystal lattice, mimic the interactions of a natural substrate with the lipidic matrix. Accordingly, the observed conformation might indeed occur in physiological conditions, with the C11 and C17 atoms of the inhibitor contacting the interface. This model, although possibly able to explain a general preference for short chain fatty acid esters, fails to account for the drop in the cutinase activity observed when increasing the length of the *only sn*-1 chain.

The total contact surface between the inhibitor and the protein is about 300 Å^2 in both molecules of the asymmetric unit. Most of the contribution is provided by the hydrophobic residues Val 184 and Leu 189 of the lipid binding site, followed by Thr 43, Ser 42 and Leu 81. Further residues in contact are Asn 84, Tyr 119, and most of the residues of the loop 180–188. In both molecules of the asymmetric unit, there are very few contacts between the residues of the substrate binding site and the region of the inhibitor mimicking the diacyl-glycerol moiety of a natural substrate. As already discussed, this lack of interactions may be related to the crystal, the inhibitor points away from the active site cleft and establishes contacts with the facing hydrophobic surface. This interaction might nevertheless be reminiscent of the physiological interaction with the lipidic interface.

The structure of the catalytic triad and of the oxyanione fole of the cut-TC4 compled is basically unchanged with respect to that of the native enzyme. As in the case of the cut-E600 complex a water molecule located close to Ser 120 in the native enzyme, is replaced by the inhibitor. The active site water molecule which favours the proper position of the Oδ2 atom of Asp 175 in the cataltic network, is conserved in spatially equivalent positions in the native and in both TC4 inhibited molecules. In both molecules of the asymmetric unit, the His 188 Nε2 atom is located 3.3 Å away from the O2 atom of the inhibitor, that is the proper spatial position which would allow the release of the alcohol and the formation of the acyl-enzyme for a natural substrate. The superimposition of the cut-TC4 (Mol B) structure onto that of the cut-E600 complex shows that the overall position of the two inhibitors in the active site is the same, thus suggesting the existence of a well-determined and favoured mode of interaction and accommodation in the active site.

The crystal structure of the cut-TC4 complex, although hardly explaining the different *sn*-1 and *sn*-2 chain length stereoselectivity of cutinase, provides a possible structural interpretation for its preference for short chain fatty acid esters. While cutinase activity on tributyrin is 40% of that of human pancreatic or RmL, on long chain triglycerides (triolein) the activity drops to 15% of the activities observed for these two lipases. The TC4 inhibitor is almost completely embedded in the active site crevice, as would be also a tributyrin molecule. Conversely, the human pancreatic lipase, which has a high activity on triolein, can accomodate a C11 mono-alkyl phosphonate inhibitor.

Cutinase activity on soluble tributyrin (below the CMC) is accounted for by the accessibility of its active site and by the deep embedding reduces the unfavourable interactions between the acyl chains of the substrate and the surrounding bulk water. At concentrations exceeding the CMC, the binding of the substrate at the lipid/water interface and its burying in the active site cleft would result in an almost complete removal of the tributyrin from the lipidic phase. Conversely, a trialein molecule, which would protrude from the cutinase active site crevice, would still continue interacting with the lipidic phase. Which situation is the most favourable? Cutinase has a lower penetrating power than pancreatic lipase or RmL in lipidic phases, as evaluated by monolayer studies. We can assume that its ability to extract, even partially, long triglycerides from the lipidic phase is therefore reduced, this leading to a lower activity. This would not be the case with true lipases which have deep and broad active sites formed upon interaction with the lipid/ water interface. It is therefore tempting to establish a correlation between the selectivity of lipases for triglycerides of certain length and the depth of their active site.

DYNAMICS OF CUTINASE

During the past few years, an increasing interest has been payed to protein dynamics as the result of the elucidation of its crucial role in protein function and enzyme catalysis. X-ray 3D structure determinations give an accurate but rather static description of crystalline proteins. Motion is deeply implicated in biological function, and the crystal structure of a protein constitutes only an average conformation. In addition, when comparing proteins in solution and in a crystalline environment, significant local differences have been found to occur, which suggests that packing forces may affect the dynamics of proteins. Crystal packing provides a stabilising environment which prevents the protein from undergoing substantial conformational changes. The X-ray resolution of various crystal forms folded in different ways makes it possible to describe protein motion by combining several static pictures. Large conformational changes have been described in this way in several protein families, such as the hexokinase, the T4 lysozyme, the transferrins and the lipases.

Although in the case of cutinase it has been demonstrated that the binding of a non sissile substrate analogue does not induce any significant structural rearrangements, minor motions can nevertheless be expected to occur in the substrate- binding region during the interaction with the substrate. No evidence in this regard can be provided by the analysis of the native and complexed forms, which constitute only the starting and final points of the interaction. The availability of several cutinase variant structures in different by means of a structural comparison among several closely related proteins in different crystal-packing environments.

TABLE 8.6: LIST OF SOLVED CUTINASE VARIANT STRUCTURES

Cutinase variant	Crystal form	Space group	Nb mol per AU	Res. (Å)	PDB code
Native (wt)	I	$P2_1$	1	1.60	1CUS
Native (wt)*	I	$P2_1$	1	1.15	1AGY
Native (wt)*	I	$P2_1$	1	1.00	1CEX
wt-EECP	I	$P2_1$	1	1.69	1XZL
wt-MUCP	I	$P2_1$	1	1.75	1XZM
S120C	I	$P2_1$	1	1.60	1CUJ
S92C	I	$P2_1$	1	1.75	1XZE
S213C*	I	$P2_1$	1	2.70	1XZD
S129C	I	$P2_1$	1	1.80	1XZA
S144C	I	$P2_1$	1	1.69	1XZF
S199C	I	$P2_1$	1	1.69	1CUU
S114Y	I	$P2_1$	1	1.75	1CUX
S189F	I	$P2_1$	1	1.69	1CUY
L81G,L182G	I	$P2_1$	1	2.10	1CUZ
N172K,R196E (I)	I	$P2_1$	1	1.75	1CUB
R17E,N172K	I	$P2_1$	1	1.75	1CUG
T45A	I	$P2_1$	1	1.69	1XZG
Y38F	I	$P2_1$	1	1.69	1XZJ
T80P	I	$P2_1$	1	1.69	1XZH
N172K	I	$P2_1$	1	1.80	1CUA
R156L	I	$P2_1$	1	1.75	1CUD
S42A	I	$P2_1$	1	1.69	1FFE

N84A	I	$P2_1$	1	1.69	1FFA
N84L	I	$P2_1$	1	1.75	1FFC
N84W	I	$P2_1$	1	1.69	1FFD
Q121L*	I	$P2_1$	1	2.70	1CUE
S120A	I	$P2_1$	1	2.70	1CUI
Y119H	I	$P2_1$	1	1.69	1XZI
R196E*	II	$P2_12_12_1$	1	1.75	1CUH
N172K, R196E (II)	II	$P2_12_12_1$	1	1.75	1CUC
N172K, R196E (III)	III	$P2_1$	3	2.70	1CUD
N84D	IV	$P2_1$	1	1.75	1FFB
A85F	V	I222	1	2.01	1CUV
wt-DFP	VI	$P2_1$	2	2.01	1XZK
wt-E600	VII	$P2_12_12_1$	1	1.90	2CUT
G82A, A85F, V1841, A185I, L189F	VIII	C2	2	2.70	1CUW
wt-TC4*	IX	$P2_1$	2	2.30	1OXM

* not included in the structural comparison; EECP: *n*-hexyl *o*-ehyl chloro phosphonate; MUCP: *n*-uncedyl *o*-methyl chloro phosphonate; DEP: di-isopropyl fluro phosphate; E600: *n*diethyl-*p*-nitrophenyl phosphate ; TC4: (R)-1,2-dibutyl-carbamoylglycero-3-O-*p*nitrophenylbutyl-phosphonate

STRUCTURAL COMPARISON

A comparative structural analysis among several cutinase variants crystallizing into 8 different crystal forms, has been carried out. Cutinase variant cystallize either in the wild-type form (type I, which is the most represented on with 26 variants) or in a form that is characteristic of the particular variant. Accordingly, two separate structural comparative analyses have been carried out, which included either the type I variants (isomorphous analysis), or one variant for each crystal form (heteromorphous analysis). For each set, the RMSD of the Cα position. In both sets, the possibility that the observed Cα fluctuations might be related either to differences in the quality of the model, or to eventual local distortions introduced by the mutation (s) or by the accomodation of the inhibitor, was checked and ruled out. Despite the overall weak flexibility detected within the set of isomorphous cutinase variants, a few discrete regions displaying a significantly higher flexibility were identified. The comparison with those of the isomorphous one, points out a much greater overall flexibility in the former case. When comparing different crystal environments, the Cα fluctuations turn out to be amplified and a few additional peaks become detectable. The stretches 24–30, 64–66, 81–87, 173–176, 180–187 and 208–211 were identified as flexible domains (Figure 8b). Asn 27 Ser 30 map on an exposed loop (residues 25–32) that is located far from and opposite to the substrate-binding region. Arg 211, the second RMSD peak, corresponds to the C-terminal residue, whereas Leu 81 and Val 184 fall into the two loops (residues 80–88 and 180–188) that delimit the active site crevice. The identification of these two stretches of residues as flexible domains provides the first direct structural evidence of the expected dynamic behaviour of the surface loops surrounding the catalytic crevice. This mobility had been previously suggested and predicted to occur in order to allow a productive interaction with the substrate. The occurrence of an appreciable flexibility in the case of Thr45, which maps on an exposed loop in the proximity of the catalytic crevice, further confirms that the structural elements that belong to the substrate-binding region have a greater mobility. As regards the location in the secondary structure elements, not surprisingly, all the flexible points fall into loops, thus indicating that the structural elements building up the protein scaffold are endow with a greater rigidity.

In both sets, the flexibility has been shown to correlate with thermal motion. In order to evaluate whether residues in intermolecular interactions has been analysed. When considering the same crystal packing environment, a reduced flexibility is observed in the case of residues/ regions greatly involved in packing contacts, and *vice versa*. However, when different crystal

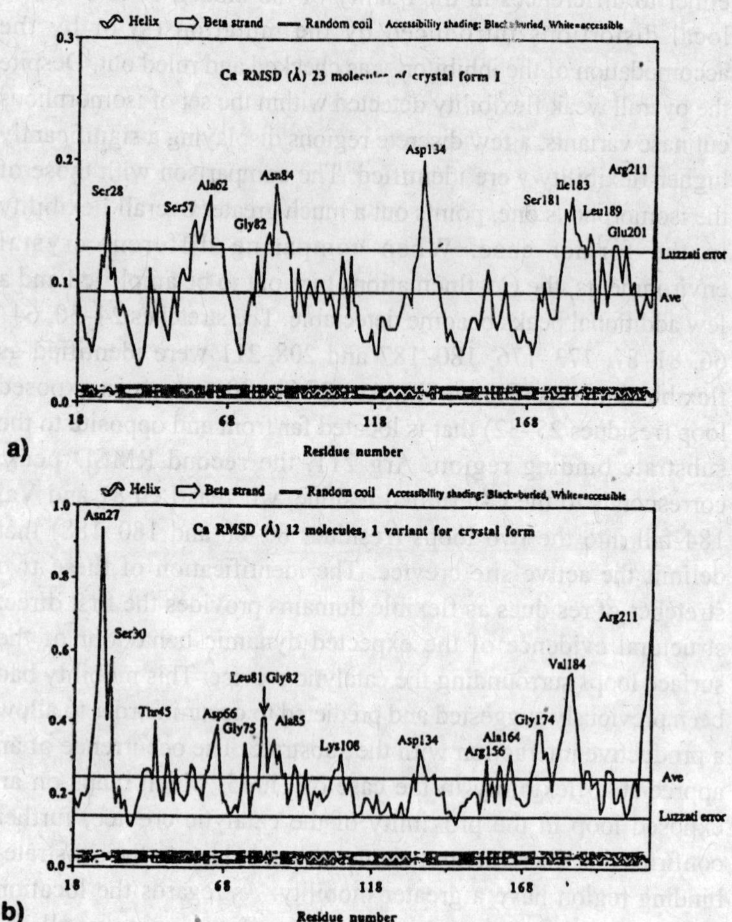

Figure 8.8 (a,b) ***Ca RMSD (Å) as a function of residue number, as obtained by including in the structural comparison (a) the 23 insomorphous variants belonging to crystal form I (isomorphous set) or (b) a single variant for each crystal form (heteromorphous set). (c, d) Secondary structure ribbon model of the Ca trace of cutinase in two different orientations. The most flexible regions detected upon heteromorphous structural analysis are highlighted. (e) Ca RMSD (Å) from the original native structure, as obtained from an 800 pa MD simulation trajectory (slid line) and as observed within the heteromorphous crystallographic data set (dotted line)***

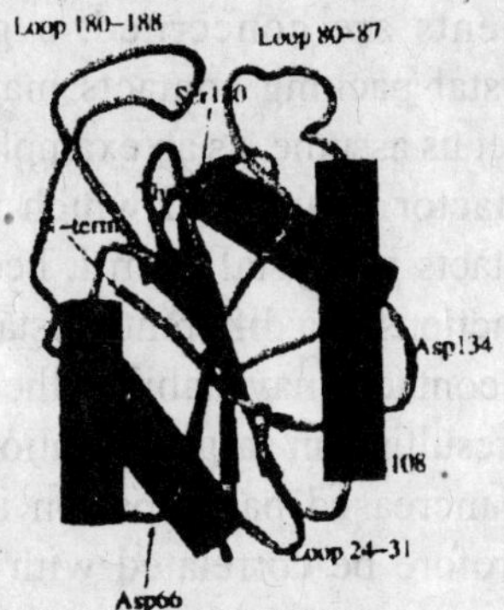
Loop 180–188
Loop 80–87
Ser120
Asp134
108
Loop 24–31
Asp66

c)

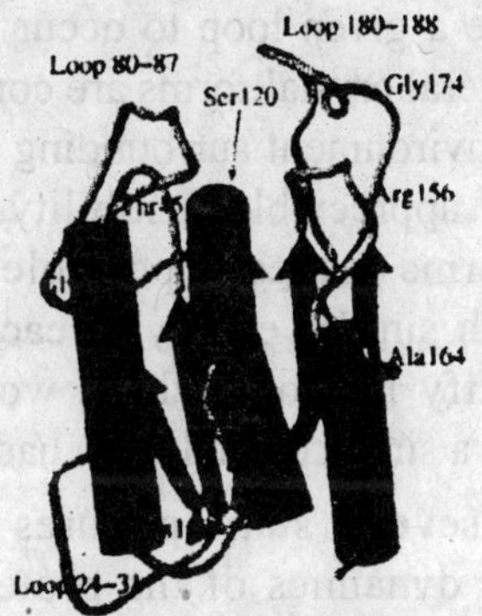
Loop 180–188
Loop 80–87
Ser120
Gly174
Arg156
Ala164
Loop 24–31

d)

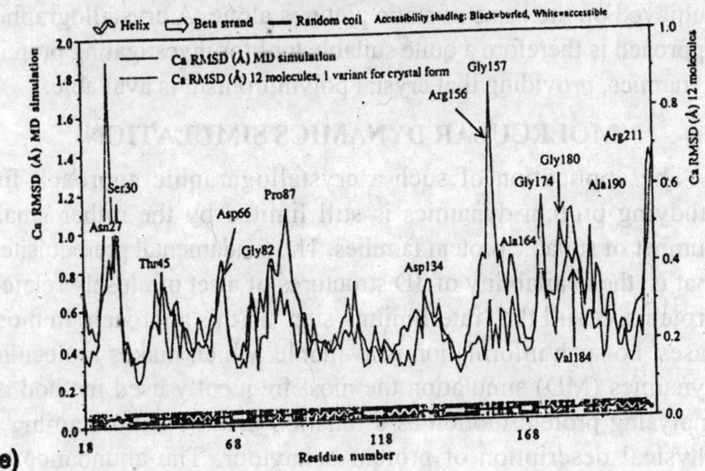
Helix
Beta strand
Random coil
Accessibility shading: Black=buried, White=accessible
Ca RMSD (Å) MD simulation
Ca RMSD (Å) 12 molecules, 1 variant for crystal form
Ca RMSD (Å) 12 molecules
Residue number
Asn27
Ser30
Thr45
Asp66
Gly82
Pro87
Asp134
Arg156
Gly157
Ala164
Gly174
Gly180
Val184
Ala190
Arg211

e)

packing environments are concerned, a greater degree of participation in crystal packing contacts may even result in a greater flexibility. Let us assume, as an example, that a given loop displaying high B-factor values and which is not involved in crystal packing contacts in crystal form I, becomes involved in intermolecular interactions in a different crystal form. The setting up of these packing contacts may stabilise the loop in a different conformation, this resulting in large deviations with respect to crystal form I. An increased participation in crystal packing contacts would therefore be correlated with greater positional fluctuations. In the same way, a reduced participation in packing contacts may cause a given loop to occur in a different spatial position. When several crystal forms are concerned, a qualitative difference in the environment surrounding a certain loop would therefore result in appreciable flexibility. The availability of several crystal forms makes it possible to detect different conformations with similar energy in each crystal form, thus allowing to identify motions which would have remained unidentified it only a single crystal form had been available.

By combining several static pictures in different crystal environments, the dynamics of the protein can therefore be inferred and described much more realistically than what can be achieved on the basis of static pictures alone. A crystallographic approach is therefore a quite suitable tool for investigating protein dynamics, providing that crystal polymorphism is available.

MOLECULAR DYNAMICS SIMULATION

The application of such a crystallogaraphic approach for studying protein dynamics is still limited by the rather small number of suitable protein families. The fundamental prerequisite., that is, the availability of 3D structures of a set of closely related proteins, is still the "rate-limiting step" in this approach. In most cases, no such information is available, which makes molecular dynamics (MD) simulation the most frequently used method of analysing protein motion as a function of time and obtaining a physical description of protein behaviour. The abundance of experimental crystallographic data on cutinase provides an excellent means of testing the reliability of MD simulation approaches.

Accordingly, the information about protein motion deduced from the comparative structural analysis have been compared with that obtained by subjecting native cutinase to a MD simulation. The Cα fluctuations calculated from the MD simulation of cutinase have been compared to the experimental mobilities observed within the heteromorphous crystallographic data set (Figure 8e). Even if an overall greater flexibility is observed upon MD simulation (as a result of the much greater degree of freedom of a protein in water rather than in a crystal environment), a fairly good agreement has been found between the experimental data , dotted line) and the MD simulation. The quite good correlation observed between the crystallographic mobilities and those inferred from a Md simulation approach is particularly relevant, since it confirms the validity of the latter approach as a means of describing protein motion.

PACKING FORCES

At present, much still remains to be elucidated about the underlying mechanism driving protein-protein interactions in the crystallization process. Only a few studies have focused on the relationships between the crystallization process. Only a few studies have focused on the relationships between the crystals. Mittl *et al*. (1994) have examined the effect of enlarging the packing contacts of crystals of glutathione reductase. They mutated two residues and thus introduced an additional crystal contact, consisting of two intermolecular hydrogen bonds. The mutations had no effect on the diffraction power or on the stability of the crystal, but the nucleation process was greatly accelerated and the crystallization time reduced. Similarly the intermolecular cross-linking of the T_4 lysozyme resulted in a more rapid crystallization of the protein analysed the molecular packing of six crystal forms of bovine pancreatic ribonuclease A obtained from different crystallization conditions. They found that the six packings differ in the number of molecules in contact and in the size of pairwise interfaces. However, they found no systematic differences in either the total area of the protein surface buried in crystal contacts or the total number of polar interactions. The protein-protein interactions in the crystal lattice turned out to be non-specific and to involve essentially all regions of the protein surface.

In the case of cutinase, most of the various crystal forms have been obtained under crystallization conditions similar to those used with the native enzyme. The availability of various crystal forms, resulting from either mutation (s) or complexation with inhibitors, provides a powerful tool to get insights into the nature of the packing forces as well as on the effect of various mutations on the crystal packing. Accordingly, the lattice contacts in the 9 different crystal forms of cutinase have been analyzed. The relationships between hydrophobicity and accessibility from one hand, and propensity to participate in crystal packing contacts on the other hand, have been investigated. With a view of determining the nature of the crystal lattice contacts, the hydrophobicities/polarities of surfaces in mutual contact have been analyzed as well.

Protein mutants with a different charge distribution form new intermolecular salt bridges or long range electrostatic interactions which are accompanied by a change in the crystal packing. The introduction or the modification of charged residues at the protein surface leads to a change in the crystal packing, so that the protein molecules become oriented in such a way as to compensate for their opposite charges. Likewise, increasing the hydrophobicity at the protein surface by means of amino acid substitutions or by complexation with hydrophobic ligands result in the establishment of new crystal packings.

The whole protein surface is involved in packing contacts and the hydrophobicities of the protein surfaces in mutual contact have been shown to be non-correlated, which indicates that the packing interactions are essentially non specific. In the case of the cydrophobic variants however, the packing contacts exhibit a certain degree of specificity, as the protein in the crystal tends to form either crystallographic or non-crystallographic dimers, which shield the hydrophobic surface from the solvent.

The likelyhood of surface atoms to be involved in a crystal contact is the same for both polar and non-polar atoms. However, when considering areas in the 200–600 Å^2 range instead of individual atoms, the surface regions either highly hydrophobic or highly polar have been found to have an increased probability to establish crystal lattice contacts. The protein surface surrounding the active site crevice of cutinase constitutes a large hydrophobic area which is involved in packing contacts in all the various crystalline contexts.

One of the most relevant implications arising from this study, is the possibility of using site-directed mutagenesis for "crystal engineering" purposes. The introduction of either charged or hydrophobic residues at the protein surface, seems to be a very promising tool to crystallize a given protein in a new form. Such an approach has been already successfully used in the case of the human H ferritin, and apoferritin.

THE ATOMIC STRUCTURE (1.0 Å) OF NATIVE CUTINASE

Implications for Protein Stereochemistry

One of the most relevant implications of atomic resolution is the possibility of a better definition of the stereochemical parameter dictionaries. To date, even though about 50% of the crystal protein structures in the Protein Data Bank are quoted as high resolution, only few of them have been solved at truly atomic resolution. Moreover, despite the increasing number of proteins for which data have been recorded at atomic resolution (for a comprehensive review see Dauter *et al.* 1995), the available data in the literature are still scarce. In most cases, small protein molecules (M_r < 10 KD_a) are concerned. Although, in the next future the number of proteins at atomic resolution is expected to exponentially increase, at present the resolution at 1.0 Å of cutinase, a 22 KDa protein, represents the largest structure ever refined to such an extent of resolution. The data were collected on the EMBL beamline X11 at the DORIS storage ring, DESY, Hamburg. The structure has been refined anisotropically to a final

R-factor of 9.4%. The overall accuracy of the final model is very high due to an observation to parameter ratio of 5.5. The mean coordinate error for protein atoms is 0.013 Å.

The availability of an accurate cutinase structure at atomic resolution allowed us to carry out an extensive analysis of several stereochemical parameters, such as peptide planarity, main chain and some side chain bond distances. The results support a need for revision of some well-established parameters in protein stereochemistry. Some of the interatomic distances in existing libraries are not accurate enough and significant discrepancies from planarity were found for the ω angles. The reliability of the observed deviations is validated by means of the associated estimated error. Large size of the proteins provides larger sample and thus more reliable estimates. The atomic structure of cutinase provides a further example which will contribute to the challenge of finding new and more appropriate parameters for protein stereochemistry.

Beyond this aspect, which is of quite general interest for structural biologists, the atomic resolution of cutinase provided insights in the catalytic mechanism and in the dynamics of this enzyme.

The Active Site Region

The X-ray data allowed the identification of most hydrogen atoms in the electron density. Accordingly, a precise and detailed description of the network of hydrogen bonds in the catalytic site has been provided. Unambiguous evidences on the protonation state of the catalytic histidine have been obtained. When an omit map was computed with the His 188 Hδ1 atom removed from the the model, a aifference Fourier peak (0 = 3) close to the Nδ1 atom could be observed . The possibility of identifying His hydrogens in the electron density has a relevant implication concerning the catalytic mechanism of cutinase: even though the occurrence of the Hδ1 atom with the concomitant absence of the Hε2 one on the catalytic histidine, had been postulated for a long time, or data represent the first direct structural evidence on the protonation state of the catalytic His of cutinase.

In the 1.0 Å structure, the Ser 120 Oγ as well as the Asn 84 Cβ, Cβ Nδ2 and Oδ1 atoms are in double conformation. The more and less occupied components of the Ser 120 Pγ are located at 2.71 Å and at 3.71 Å from the His 188 Nε2 atom, respectively (Figure 9b). The less occupied position, being located too far away from the catalytic histidine, is not likely to play a relevant role in catalysis. The active site water molecule has been modelled as two sites, in which the occupancy of the less occupied member was tide to that of the less occupied member of the Ser 120 Oγ. The two Ser 120 side chain conformations are hydrogen bonded to the two components of the solvent molecule in double conformations are hydrogen bonded to the two components of the solvent molecule in double conformation in an arrangement scheme in which the less occupied member of the solvent molecule faces the more occupied Ser 120 Oγ conformation . The third component of the oxyanion hole, i.e. the Ser 42 Oγ atom, is stabilised by two hydrogen bonds with the Gln 121 Nε2 atom (3.15 Å) and with the more occupied Nδ2 site of the Asn 84 side chain (3,06 Å). The less occupied Asn 84 Nδ2 site, occurring at 3.96 Å from the Ser 42 Oγ, does not seem to contribute to the stabilization of the Ser 42 hydroxyl group position. The less occupied alternate conformation of Asn 84 is stabilised by the solvent site in double conformation. The Ser 42 side chain displays an appreciable extent of flexibility, as indicated by the electron density map which points out a high extent of anisotropy.

The atomic resolution allowed to detect an appreciable extent of flexibility in the active site of cutinase. Although the less occupied alternate conformations do not seem to play a role in catalysis, they clearly indicated that the active site is less rigid than suggested by previous crystallographic studies. These results are in agreement with those recently obtained by an NMR study of cutinase which showed that the residues in, or near, the active site display functional motions in the ms-ms time scale. The biological relevance of this flexibility might reside in a possible adaptation to different substrates which might be amplified in the course of evolution.

THE ANALYSIS OF THE ANISOTROPY

The refinement of cutinase at atomic resolution, beyond the possibility of detecting more sensibly atomic motions, provide information into their anisotropic nature. In order to get some quantitative insights into the distribution of anisotropy within the protein, the anisotropy (A) of each atom was expressed as

$$A = \sigma(\lambda_i) / < \lambda_i >$$

where λ_1, λ_2 and λ_3 are the principle components of the U_{il} matrix, $<\lambda_\iota>$ and σ (λ_ι) are the mean and the standard deviation of λ_ι, respectively. The computation method chosen has the advantage of rendering A independent from thermal motion, thus providing a convenient measure of the "ellipsoidicity". The average anisotropy of each residue has been computed. A few highly anisotropic regions (25–31, 62–67 and 134–136) and residues can be observed. The region exhibiting the greatest anisotropy is the loop 24–31. As already mentioned, this loop, which is exposed to the solvent and opposite to the substrate binding site, displays a significantly greater extent of thermal motion in all the different crystal forms of cutinase variants solved so far. Besides this disordered loop, the loops 80–88 and 180–188 exhibit an extent of anisotropy greater than the average. In Figure are shown the 25% thermal ellipsoids of the loop 80–88 which, together with the loop 180–188, builds up the catalytic crevice. Based on the visual inspection, a main coherent direction for motion can be identified which runs from the left bottom to the right top corner of the figure. This motion, bringing the loop 80–88 close to the facing loop 180–188, produces a sort of breath-like movement in the substrate binding region. The high extent of thermal motion and of anisotropy of this loop is an agreement with what expected for an efficient substrate binding region, which should display a certain flexibility in order to allow productive interaction with the substrate. The analysis of the anisotropic motions confirms the indications previously obtained on the dynamic behaviour of the substrate binding site of cutinase. **Conclusions and Perspectives**

The availability of an atomic model of cutinase and of the 3D structure of several cutinase mutants and covalently inhabited complexes, have greatly contributed to elucidate the structure-function relationships of cutinase. Despite this enormous body of information, the precise behaviour of cutinase *during* the interaction with the substrate, i.e. the mode of approach to the interface, and the release of the substrate still remain to be elucidated. The structure of cutinase covalently complexed with an analogue of triglyceride has revealed that the acyl moiety of the ligand establish very few interactions with the substrate binding region, thus failing to provide a convenient explanation for the stereoselectivity of cutinase. Some elucidations in this regard are likely to be provided by time-resolved crystallographic approaches, which should allow to follow the release of the substrate. In this frameship, the accuracy of structural information obtained from very rapid *Laue* data collection (X-ray pulses of 150 psec and total exposure time shorter than 10 nsec) on native cutinase, assumes a particular relevance and is very encouraging. Since tertiary structural changes in proteins general occur in the msec to nsec timescale at physiological temperatures, new opportunities are opened in the field of time resolved crystallography. Short laser pulses could be used to trigger a photoreaction in cutinase crystals by using suitable caged compounds. The structural modifications intervening during the release of the substrate could be probed at various time delays (in the msec to nsec time scale) by recording Laüe data from single X-ray flashes.

From an applicative standpoint, the identification of the structural determinants responsible for the specificity and stereoselectivity of cutinase may open the way to the rational design of point mutants endowed with a different substrate preference. New variant forms could therefore be obtained by site directed mutagenesis approaches, thus allowing the use of cutinase in the catalysis of new substrates. Cutinase variants with an increased resistence to detergents and high temperatures have been already obtained and are presently used in the detergent industry.

9

Genetics, Structure and Applications of Lipases

Fungi of the genus *Mucorales* are well-established producers of useful enzymes, notably proteases and lipases. Some of these enzymes have found important commercial application, e.g. the protease from *Rhizomucor miehei* which is used as a microbial rennet in milk clotting. The lipase from *Rhizomucor miehei* which is commercially available in immobilized form as LIPOZYME^{TM}, has also been studied in a wide range of technical applications. Compared to this commercial success, lipases of the genus *Rhizopus* have been less successful, though they are also commercially available. For instance, the Japanese enzyme producer Amano Co. in Nagoya provides lipase N, D and J which are derived from *F. niveus, R. delemar* and *R. javanicus*, respectively. Lipases from *Rhizomucor miehei* and *Rhizopus* show 1,3 regiospecificity toward tricaylglycerols but exhibit slightly different stereoselectivity and chain length specificity. A recent reclassification of these fungi had *R. niveus, R. delemar* and *R. javanicus* all renamed as *Rhizopus oryzae* and, consistent with this taxonomic evaluation, the lipases isolated from *R. delemar* (RDL), *R. javanicus* (RJL) and *R. niveus* (RNL) were found to have identical amino acid sequences; the lipase from *R. oryzae* (RPL) differs by only two conservative substitutions. Thus, the different properties of these lipases observed in commercial preparations may be due to proteolytic cleavage of the same prolipase, which contains extra amino acid residues to guide folding and secretion, at different amino acid residues.

TABLE 9.1: SOME PROPERTIES OF *MUCORALES* LIPASES

Lipase		pI	pH optimum	Temp-erature optimum (°C)	Specific activity (U/mg)*
Rhizopus oryzae (ROL)	recombinant[1)]	8.3[2)]	8.0–8.5	30	10000
	native[3)]	n.d.	8.5	40	10000
Rhizopus delemar (RDL)	native[4)]	8.6	8.0–8.5	30	5094
	recombinant[5)]	8.1[2)]	8.0	30	n.d.
Rhizopus niveus (RNL)	native[6)]	n.d.	6.0	40	n.d.
Rhizopus javanicus (RJL)	native[7)]	7.1, 7.7, 7.8	6.0–8.0	n.d.	9260
Rhizomucor miehi (RML)	native[8)]	3.9	7.0	n.d.	7500
	recombinant[9)]	n.d.	n.d.	70	8810

1) Beer *et al.* (1997), 2) deduced from amino acid sequence, 3) Salah *et al.* (1994), 4) Haas *et al,* (1991), 5) Joerger and Haas (1993), 6) Kohno *et al.* (1994), 7) Uyttenbroeck *et al.* (1993), 8) Huge-Jensen *et al.* (1987), 9) product sheet, Novo Nordisk; *under different assay conditions

The structure and function of both *Rhizomucor miehei* lipase and several *rhizopus* lipases has been investigated on a molecular basis using x-ray crystallography, computer-modelling, site-directed mutagenesis and kinetic experiments with monolayers and pseudosubstrates. In the framework of these and other studies, these enzymes have been cloned, expressed in heterologous hosts and genetically modified. As a result, *Rhizopus* lipases today are a class of enzymes whose function is reasonably well understood. As they were found already to have a strong potential is reasonably well understood. As they were found already to have a strong potential in a wide range of reactions relevant to industry, the better understanding of their structure and function now achieved should be considered as a promising step towards their improvement *via* protein engineering and their wide application in industrial enzyme technology.

BIOCHEMICAL PROPERTIES

The regio- and stereoselectivity of pure *Rhizopus* lipases, their pH-and temperature optimum are summarized in Table 1 and compared to the lipase from Rhizomucor *miehei.* All enzymes are highly 1, 3-specific in the hydrolysis or esterification of glycerides. They are most active at neutral or slightly basic pH and at 30°C and exhibit a moderate pH-and temperature stability. If immobilized, the lipase of RML became stable up to 70°C.

The native enzymes are only moderately glycosylated, e.g. 11% carbohydrate content of native RML. As fully active recombinant lipases can be obtained from heterologous expression in *E. coli*, glycosylation does not seem to be necessary for lipase activity.

MOLECULAR GENETICS

Cloning and Nucleotide Sequence

Up to now, the genes of four *Mucorales* lipases have been cloned, and three have been over expressed and purified (Table 1). Lipase genes from *Rhizopus delemar* ATCC34612 (RDL), *Rhizopus niveus* IFO 9759 (RNL), *Rhizopus oryzae* ATCC 853 (ROL) and *Rhizomucor miehei* (RML) were cloned from cDNA libraries screened either by probes derived from the amino acid sequences or by detecting lipase-producing *E. coli* transformation on tributyrin plates. Analysis of the nucleotide sequence and comparison with the N-terminal sequences derived from the native lipase revealed that all *Mucorales* lipases studied of far are organized as preproenzymes. From the amino acid sequence, a molecular weight of 42 kDa and 39.5 kDa for the lipase precursors of ROL and RMA was deduced, and 29.5 kDa for the mature ROL and RML lipases.

At the amino acid level, RNL and RDL are identical, whereas ROL differs in two amino acids in the mature lipase and in six of the prepropeptide. The amino acid sequence of RML, if compared to ROL, shows 54% identity of the mature lipases and 29% identity of the prepropeptides. While *Rhizopus delemar* and *Rhizopus niveus* have been reclassified as *Rhizopus oryzae*, comparison of the nucleotide sequence of the three lipases genes indicates, that RDL and RNL are identical, but differ slightly from ROL.

Figure 9.1: ***Organization of the lipases from* Rhizopus oryzae *(ROL) and* Rhizomucor miehei *(RML) as preproenzymes. The signal sequence (white box) is followed by the propeptide (gray box) and the mature lipase (black box). The numbers on top display the cleavage sites and the total size of the protein. Numbers in brackets indicate an alternative cleavate sites and the total size of the protein. Numbers in brackets indicate an alternative cleavage site observed with the native lipase of* Rhizopus niveus**

Overexpression and Purification

cDNA-derived clones of *E. coli* expressing RML, RNL and ROL showed low lipolytic activity on tributyrin plates. Under the control of the lac-promotor on the pUC8- 2.14 cloning vector, a low-level expression of RDL in *E. coli* was achieved. Visualization of the expressed lipase on SDS-gels by immunoblotting resulted in bands with sizes of 45 and 39 kDa, corresponding to the proper- and proenzyme, respectively. Similar results were found for recombinant ROL, indicating that *E. coli* is unable to fully process the preproenzyme to mature product.

The genes encoding the ROL and RDL were therefore cloned into high-level expression vectors, in order to allow for the production of larger amounts of recombinant lipase in *E. coli*. It turned out, however, that the expression of even low amounts of active mature lipase upon translocation into the periplasmic space low amounts of active mature lipase upon translocation into the periplasmic space is toxic for the *E. coli* cells. Expression of mature lipase was possible only if a tightly controlled promoter was used, or cytoplasmically, which resulted in expression of mature inactive lipase as inclusion bodies. However, expression of the active prolipase into the periplasm was not toxic for *E. coli*. Using a pET-vector, both the mature lipase and prolipase of

Rhizopus delemar were cytoplamatically expressed in *E. coli* under the control of the T7- promotor at a level of 15–20% as inactive inclusion bodies.

Instead of the less tightly controlled pET-vector system, in our group of the *E. coli* expression vector pCYTEXP1 for the expression of ROL under the control of the strictly regulated 1 P_RP_L promoter was chosen. Both mature and precursor lipase were fused to an ompA-leader sequence, which directs the expression of the lipase into the periplasm. Using this system, Expression levels of 20–30% and 10-20% were obtained for the mature and prolipase, respectively. The mature lipase, which was expressed to a higher level, formed inactive inclusion bodies in the periplasm, whereas the prolipase was partly active.

For the inclusion bodies containing mature and precursor ROL and RDL, a refolding and purification protocol was established, leading to specific activities of 10000 U/mg (ROL) and 5094 U/mg (RDL) for the mature lipases, and of 12800 U/mg (ROL) and 5800 U/mg (RML) for the prolipases. Thus, the specific activity obtained with the recombinant mature ROL was almost twice as high compared to the recombinant RDL, and 20% higher compared to the native lipase of *Rhizopus delemar*, but similar to that of the native ROL. Recently, expression of functional ROL in high yield and purity could be achieved using *Pichia pastoris* or *Saccharomyces cerevisiae* as a host. RML could be functionally expressed at high level by cloning the gene encoding for the lipase precursor under the control of the α-amylase promoter of *Aspergillus oryzae* and transforming *Aspergillus oryzae* with the resulting plasmid. The secreted lipase was correctly processed in *Aspergillus oryzae*, and purification led to a recombinant lipase with a specific activity of 8810 U/mg as compared to 7500 U/mg of the native lipase. It thus appears that fungal expression systems are better suited than *E. coli* to express functional lipases of the genus *Mucorales* at a high level.

THE FUNCTION OF THE ROL PROSEQUENCE

All *Mucorales* lipases investigated so far have a prosequence, which is unusual for microbial lipases. Genetic studies on the role of this 97 amino acids prosequence suggest at least two functions. First, it allows secretion of ROL without damaging the cells. Thus, expression studies of ROL and Pro ROL in *E. coli* showed that even low levels of actively expressed ROL in the periplasm of *E. coli* showed that even low levels of actively expressed ROL in the periplasm of *E. coli* is toxic for the cells, whereas Pro ROL is not. The main reason for the toxicity might be the fact, that ROL, in contrast to ProRPL, possesses phospholipase activity. Thus, the function of the prosequence is analogous to that of other zymogen sequences.

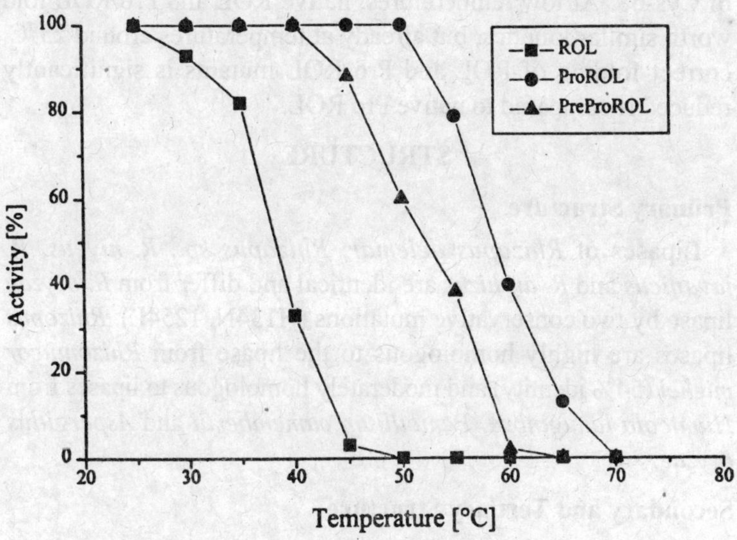

Figure 9.2: ***Activity of ROL, Pro ROL and PrePro ROL as a function of temperature***

Secondly, it seems to play an important role in the folding of the lipase, particularly in the correct formation of disulphide bridges. Both Pro ROL and ROL could be refolded and purified to a similar specific activity of 12800 and 10000 U/mg, respectively, demonstrating that Pro ROL is fully active and thus does not behave like a typical zymogen. However, the enzymatic properties of both lipase forms differed markedly. ROL had a significantly broader and higher pH optimum (pH 8.5) in comparison to Pro ROL (pH 7.5). Additionally, Pro ROL was thermostable up to 50°C, whereas ROL was stable only up to 30°C . A difference between ROL and Pro ROL was also observed in terms of substrate affinity, suggesting that structural elements other than the lid are involved in interfacial activation. By systematic mutation analysis of the resulting Pro ROL and ROL mutants under different conditions, it was shown that the prosequence facilitates folding via a pathway including Cys-68 as the key residue, acting as an intramolecular thiodisulphide reagent during folding. Furthermore, it was shown that residues in the neighbourhood of Cys-68 play an important structural role, in particular Arg 69 which enhances the leaving group character of Cys-68. At low temperatures, native ROL and Pro ROL fold worth similar kinetics, but already at temperatures around 25°C, correct folding of ROL and Pro ROL mutants is significantly reduced if compared to native Pro ROL.

STRUCTURE

Primary Structure

Lipases of *Rhizopus delemar, Rhizopus sp., R. niveus, R. javanicus* and *R. arrhizus* are identical and differ from *R. oryzae* lipase by two conservative mutations (H134N, 1254L). *Rhizopus* lipases are highly homologous to the lipase from *Rhizomucor miehei* (54% identity) and moderately homologous to lipases from *Humicola lanuginosa, Penicillium camembertii* and *Aspergillus oryzae.*

Secondary and Tertiary Structure

For four *Mucorales* lipases the structure has been determined in closed and open form including two complexes with substrate analogous inhibitors.

Lipases are members of the α/β hydrolase fold family a central hydrophobic eight-stranded mixed β-pleated sheet with the β-strands connected by loops and amphiphilic α-helices. Secondary structure is well conserved among the mucorales group and homologous lipase, but is slightly mediated by crystallization conditions. Upon opening of the lid, the helix increases in length.

```
R.delemar     -122 MVSFISISQGVSLCLLVSSMMLGSSAVPVSGKSGSSNTAVSASDNAALPPLISSR
R.miehei       -93 --MVLKQRANYLGFLIVFFTAFLVEAVPIKRQSNS--------TVDSLPPLIPSR
P.camembertii  -25 ------MRLS--FFTALSAVASLGYALPGKLQSR---------------------
H.lanuginosa       -------------------------------------------------------
A.oryzae       -27 ------MRFLSGFVSVLSSVALLGYAYPTAIDVR---------------------

R.delemar      -67 CAPPSNKGSKSDLQAEPYNMQKNTEWYESHGGNLTSIGKRDDNLVGGMTLDLPSD
R.miehei       -48 TSAPSSSPSTTDPEAPAMSRNG---------------PLPSDVETKYGMALNATSY
P.camembertii      -------------------------------------------------------
H.lanuginosa       -------------------------------------------------------
A.oryzae           -------------------------------------------------------

R.delemar      -12 APPISLSSSTNSA|SDGGKVVAATTAQIQEFTKYAGIAATAyCRSVVPGNKWDCVQ
    A                                  EEE  HHHHHHHHHHHHHHHHHHHH
    B                                       HHHHHHHHHHHHHHHHH               HH
  R.niveus                             EEE  HHHHHHHHHHHHHHHHHH               HH
R.miehei        -7 PDSVVQAM-----|SIDGGIRAATSQEINELTYYTTLSANSYCRTVIPGATWDCIH
   closed                              EEE  HHHHHHHHHHHHHHHHHH               HH
   open                                EEE  HHHHHHHHHHHHHHHHHH
P.camembertii    1 -------------+-------DVSTSELDQFEFWVQYAAASYYEADYTAQVGDKLS
                                            HHHHHHHHHHHHHHHHHHH  HHHHH
H.lanuginosa     1 -------------+-------EVSQDLFNQFNLFAQYSAAAYCGKNNDAPAGTNIT
                                            HHHHHHHHHHHHHHHHHH   HHH
A.oryzae         1 -------------+-------DIPTTQLEDFKFWVQYAAATYCPNNYVAKDGEKLN
                                                  .       .*  .*       .     .

R.delemar       43 CQK------WVPDGKIITTFT-SLLSDTNGYVLRSDKQKTIYLVFRGtNSFRSaI
    A                             EEE            EEEEE           EEEE     HHHHH
    B              HHH            EEE                 EE          EE       HHHH
  R.niveus         HHH       H    EEEEEEE  E    EEEEEEEE     EEEEEEE       HHHH
R.miehei        43 CD-------ATEDLKIIKTWS-TLIYDTNAMVARGDSEKTIYIVFRGsSSIRNWI
   closed          HH             EEEEEEE        EEEEEEE     EEEEEE       HHHHH
   open                           EEEEEEE        EEEEEEE     EEEEEE       HHHHH
P.camembertii   36 CSKGNCPEVEATGATVSYDFSDSTITDTAGYIAVDHTNSAVVLAFRGsYSVRNWV
                          HHHHH   EEEEE             EEEEE    EEEEEE       HHHHH
H.lanuginosa    36 CTGNACPEVEKADATFLYSFEDSGVGDVTGFLALDNTNKLIVLSFRGsRSIeNwI
                    HHH   HHHHH   EEEEEE        EEEEEEEE     EEEEEE        HHHH
A.oryzae        36 CSVGNCPDVEAAGSTVKLSFSDDTITDTAGFVAVDNTNKAIVVAFRGsYSIRNWV
                   *                          .*  . .           . . ***.*     .

R.delemar       91 TdIvfNFSDY-KPVKGAKVHAGfLSsYEQVVNDYFPVVQEQLTAHPTYKVIVTGh
    A                     EEE        EEEHHHHHHHHHHHHHHHHHHHHHHHHHHH     EEEEEE
    B                                     HHHHHHHHHHHHHHHHHHHHHHHHHH      EEEEE
  R.niveus         H      EEE        EEEHHHHHHHHHHHHHHHHHHHHHHHHHHHH   EEEEEEE
R.miehei        90 ADLTFVPVSY-PPVSGTKVHKGFLDSYGEVQNELVATVLDQFKQYPSYKVAVTGH
   closed          HH     EEE        EEEHHHHHHHHHHHHHHHHHHHHHHHHHHHH   EEEEEEE
   open            HHH    EEE        EEEHHHHHHHH   HHHHHHHHHHHHHH      EEEEEEE
P.camembertii   91 ADATFVHTNP-GLCDGCLAELGFWSSWKLVRDDIIKELKEVVAQNPNYELVVVGH
                   HH     EE          EEHHHHHHHH   HHHHHHHHHHHHHHHH    EEEEEE
H.lanuginosa    91 GNLNFDLKEINDICSGCRGHDGFTSSWRSVADTLRQKVEDAVREHPDYRVVFTGH
                          EEE        EEEHHHHHHHHHHHHHHHHHHHHHHHHHHHH   EEEEEEE
A.oryzae        91 TDATFPQTDP-GLCDGCKAELGFWTAWKVVRDRIIKTLDELKPEHSDYKIVVVGH
                   ..   *          *      **  .    * .          . .   * .   **
```

```
R.delemar        145 SLGGAQALLAGMDLYQREPRLSPKNLSIFTVGGPRVGNPTFAYYVEST--GIPFQ
    A                E HHHHHHHHHHHHHHH        EEEEEE       HHHHHHHHH      EE
    B                 HHHHHHHHHHHH         HHH EEEEE       HHHHHHHHH      EE
  R.niveus            HHHHHHHHHHHHHHHH        EEEEEE       HHHHHHHHHH     EE
R.miehei         144 SLGGATALLCALDLYQREEGLSSSNLFLYTQGQPRVGDPAFANYVVST--GIPYR
    closed           HHHHHHHHHHHHHHHH         EEEEEE       HHHHHHHHH      EE
    open             HHHHHHHHHHHHHHHH         EEEEEE       HHHHHHHHH      EE
P.camembertii    145 SLGAAVATLAATDLRG--KGYP--SAKLYAYASPRVGNAALAKYITAQ--GNNFR
                     HHHHHHHHHHHHH             EEE         HHHHHHHHH
H.lanuginosa     146 SLGGALATVAGADLRG--NGY---DIDVFSYGAPRVGNRAFAEFLTVQTGGTLYR
                      HHHHHHHHHHHHH           EEEEEE       HHHHHHHH      EEE
A.oryzae         145 SLGAAIASLAAADLRT--KNY---DAILYAYAAPRVANKPLAEFITNQ--GNNYR
                     ***.* * . . **             ... . ***..   * ..      *  .

R.delemar        198 RTVHKRdIVPHvPPQSfGFLHPGVESWIKSGTS-NVQICTSEIETKDCSNSIVP-
    A                EEEE   HHHH             EEEE       EEEE         HHHH
    B                EEE     HHH   HHH      EEEE         EE
  R.niveus           EEEE    HHH   HHH      EEEEEEE    EEEEE         HHH
R.miehei         197 RTVNERDIVPHLPPAAFGFLHAGEEYWITDNSPETVQVCTSDLETSDCSNSIVP-
    closed           EEEE    HHH   HHH      EEEEEEEE  EEEEEE         HHHH
    open             EEEE    HHH   HHH      EEEEEEEE  EEEEEE         HHHH
P.camembertii    194 FTHTN-DPVPKLPLLSMGYVHVSPEYWITSPNN--ATVSTSDIKVIDGDVSFDGN
                             HHH   HHH                      HHH           HH
H.lanuginosa     196 ITHTN-DIVPRLPPREFGYSHSSPEYWIKSGTL--VPVTRNDIVKIEGIDATGGN
                     EEE     HHH   HHH       EEEE           HHHEEEE
A.oryzae         193 FTHND-DPVPKLPLLTMGYVHISPEYYITAPDN--TTVTDNQVTVLDGYVNFKGN
                     .*    * **..*    *. *   *  *         .    .   .

R.delemar        251 ---------FTSILDHLSYFdINeGSCL---------
    A                              HHHEE  EE
    B                            HHHH
  R.niveus                        HHHH    EE
R.miehei         251 ---------FTSVLDHLSYFGINTGLCT---------
    closed                        HHHH
    open                          HHHH
P.camembertii    246 TGT--GLPLLTDFEAHIWYFVQ-VDAGKGPGLPFKRV
                     H           HHH
H.lanuginosa     248 NQP-----NIPDIPAHLWYFGL-IGTCL---------
                                  HHHH
A.oryzae         245 TGTSGGLPDLLAFHSHVWYFIH-ADACKGPGLPLR--
                                    *. **
```

Figure 9.3: ***Sequence alignment of lipases. The start of the nature lipase is indicated by a vertical bar (signal and propeptide left of thye bar). R. oryzae lipase is nearly indentical to R. Delemar lipase (H134N and 1254L). In bold: catalytic triad, oxyanion hole, mutation sites (in lower case letters); the lid helix is 185-D91 in RML; secondary structure analysis by DSSP (Kabsch and Sander, 1983), alignment by CLUSTAL W (Thompson et al., 1994), visulaization by BOXSHADE (shareware program by Kay Hofmann, Lausanne)***

In the lipases of the *Mucorales* group, the tertiary structure is further stabilized by three disulfide bridges. The catalytic triad is placed on top of a β-sheet, with the catalytic serine and aspartic acid in two separate loops between a β-strand and an α-helix and the catalytic histidine in a short α-helix near the C-terminus of the lipases. The position of these residues and the oxyanion hole are highly conserved. As in most lipases, a mobile lid covers the catalytic site in the catalytically inactive state. In *Mucorales* lipases, the lid consists of a short α-helix (4–8 amino acids) linked to the body of the lipase by flexible structure elements. In the open, active form of the lipase, the lid moves away and renders the substrate binding site accessible. It has been suggested, however, that lid opening in *Mucorales* lipases does not follow a simple two-state model, but more conformations are possible, some of whom have been observed depending on the crystallization conditions.

TABLE 9.2: *MUCORALES* LIPASE STRUCTURES PUBLISHED IN PROTEIN DATA BANK (PDB) OR MODELLED BY HOMOLOGY.

Lipase source	PDB entry	Confor-mation	Inhibitor	Resolution	Reference
R. delemar	1TIC A,B	semi-open	none	2.6 Å	Derewenda *et al.* (1994b)
R. niveus	1LGY A,B,C	closed	none	2.2 Å	Kohno *et al.* (1996)
R. miehei	1TGL	closed	none	1.9 Å	Brady *et al.* (1990)
”	3TGL	Closed	none	1.9 Å	Derewenda *et al.* (1992a)
”	4TGL	open	diethylphosphate	2.6 Å	Derewenda *et al.* (1992b)
”	5TGL	open	hyxylphosphonate-	3.0 Å	Brzozowski *et al.* (1991)
R. oryzae	model	open	Beer *et al.*(1996b)		
P. camembertii	1TIA	closed	none	2.0 Å	Derewenda *et al.* (1994)a
H. lanuginosa	1TIB	closed	none	1.8 Å	Derewenda *et al.* (1994b)

The hydrolytic reaction pathway of lipases is similar to that of serine proteases and consists of 5 subsequent steps: *i*) the ester substrate binds, ii) a first tetrahedral intermediate is formed by nucleophilic attack of the catalytic serine, with the oxyanion stabilized by two or three hydrogen bonds, the so-called oxyanion hole, iii) the ester bond is cleaved, and the alcohol moiety leaves the enzyme, iv) water adds to the acyl enzyme forming a second tetrahedral intermediate, v) the acyl enzyme is cleaved.

Binding of phosphate and phosphonate inhibitors mimicks the first tetrahedral intermediate state of substrates. From the x-ray structure of complexes of *R. miehei* lipase with diethylphosphate and hexylphosphonate ethylester, the binding sites for the alcohol

Figure 9.4: ***Secondary structure of*** **R. miehei** ***lipase in closed (left) and open (right) form***

and the fatty acid moiety were identified. Like in other lipases, the alcohol binding site consists of two binding pockets for the large and the medium sized substituent of a secondary alcohol, which mediate stereoselectivity towards esters of secondary alcohol. The residues which mediate binding of the scissile fatty acids can be extracted from the experimentally determined structure,. The fatty acid binding site is formed by a hydrophobic crevice, which extends beyond the inhibitor. The crevice is long enough to allow binding of fatty acids up to a length of C18. While the scissile fatty acid binding site is easily identified, no x-ray structure is available where the binding site of the second and third fatty acid of a triglyceride substrate is indicated. From an analysis of surface hydrophobicity a second, shallow hydrophobic binding site was postulated which might be able to bind a fatty acid. Modelling of substrate binding to these binding sites has been used to understand fatty acid chain length and stereoselectivity towards triglycerides and analogs.

Although the five lipases described here have only moderate overall homology, most of the amino acids involved in catalysis are conserved, with two exceptions:

(1) in *Rhizopus* lipases, the oxyanion hole is formed by Thr83, and by serine in the other lipases, and the binding pocket for medium-sized substituent of secondary alcohol, Leu258 of *R. delemar* lipase which is conserved in the lipases of *R. miehei* and *H. lanuginosa,* and Ile and Val in *P. camembertii*

TABLE 9.3: MUTANTS OF FUNCTIONALLY RELEVANT AMINO ACIDS

Lipase source	Mutation (%)	Role of the mutated residue	Effect[a]	References
R.oryzae	Y28F	anchor between active site serine etrahedral substrate intermediate	0.02	Beer *et al.* (1996a)
R.oryzae	T83A	stabilization of tetrahedral substrate intermediate	0.04	Beer *et al.* (1996a)
R.oryzae	T83S	"	12	Beer *et al.* (1996a)
H. lanuginsa	E87A	defines lid position	35–70	Martinelle *et al.* (1996)
R. oryzae	A89W	A in RDL, W in RML	56	Beer *et al.* (1996a)
R. oryzae	D92N	hydrogen bond to T83	7	Beer *et al.* (1996a)
R. delemar	S115D, L146K	near the active site	0	Klein *et al.* 1997
R. oryzae	H144F	maintains active site geometry by H-bridging between Y28 and H257 carbonyl	0.04	Beer *et al.* (1996a)
R. oryzae	H144S	"	2.2	Beer *et al.* (1996a)
R. oryzae	D204A	catalytic residue	0.06	Beer *et al.* (1996a)
R.oryzae	D262G	not involved in hydrolysis	96	Beer *et al.* (1996a)
R. oryzae	E265Q	not a catalytic residue, but maintains active site geometry	98	Beer *et al.* (1996a)
R.oryzae	E265D	maintains active site geometry	0.14	Beer *et al.* (1996a)

[a]Change of activity relative to wild-type

and *A. oryzae* lipases, respectively;

(2) the sequence of the lid helix differs considerably.

The residues mediating substrate specificity differ to a bigger extent. It has been suggested to relate these differences to the different biochemical properties of these lipases concerning their activity, stereoselectivity, stability and substrate specificity. Lipases from *R. delemar, R. miehei* and *H. lanuginosa* are triacylglycerol hydrolases, *A. oryzae* lipase is a diacylglycerol hydrolase and *P. camembertii* lipases is a mono- and diacylglycerol hydrolase.

PROTEIN ENGINEERING OF *RHIZOPUS* LIPASES

Site-directed mutagenesis has been applied to identify the function of individual amino acids in *Rhizopus* lipases. Modelling of substrate binding and enzymatic catalysis helped to understand their function and to predict mutants with defined properties. Two groups of mutants have been predicted and characterized: mutants related to the enzymatic function and mutants with altered substrate specificity.

In addition to the catalytic triad, a network of other amino acids near the active site are necessary to maintain the proper orientation of catalytic residues in *R. oryzae* lipase. Furthermore, mutation of Thr83 to Ser, Ala and Val indicated that the side chain of this residue is involved in stabilization of the oxyanion, and that formation of the acyl enzyme is the rate limiting step in ester hydrolysis.

Numerous experiments have been performed to probe the fatty acid binding site and to change fatty acid specificity both near and far away from the active site. The effect of these latter mutations can be easily understood: increasing the size or the polarity of side chains lining the scissile binding site prevents long-chain fatty acid chains reaching the mutated site from binding, while binding of shorter fatty acid chains is unimpaired. The observed role of the side chain of Thr83, forming the oxyanion hole, for fatty acid chain length specificity is, however,

TABLE 9.4: MUTANTS WHICH CHANGE FATTY ACID SPECIFICITY

Lipase source	Mutations	Role of the mutated residue	Effect of the mutation on the hydrolytic activity	Reference
R. delemar	T83S	oxyanion hole; binding of acyl chain	4 fold decrease of tributyrin, 20% decrease of tricaprylin relative to triolein	Joerger and Haas (1994)
H. lanuginosa	W89 F,L.G.E	binding of acyl chain	Increase of triacetin and trioctanoin relative to tributyrin	Martinelle *et al.* (1996)
R. delemar	V94W	binding of fatty acids >C8	1.4-fold increase of tricaprylin relative to tributyrin	Klein *et al* (1997)
R.oryzae	F95Y	binding of fatty acids >C12	60% (30%) increase of caproic acid methyl ester relative to oleic (stearic) acid	Atomi *et al.* (1996)
R. dlemar	F95D	binding of fatty acids	2-fold decrease of hydrolysis	Joerger and Haas (1994)
R. delemar	F95D F214R	distal end of acyl binding groove	3-fold increase of tricaprylin relative	Klein *et al.* (1997)

Lipase source	Mutations	Role of the mutated residue	Effect of the mutation on the hydrolytic activity	Reference
			to triolein	
R. delemar	F112W	fatty acid binding	50% increase of tributyrin relative to triolein	Joerger and Haas (1994)
R. delemar	F112Q	fatty acid binding	no activity	Klein *et al*. (1997)
R.delemar	V206T	fatty acid binding	10–20% activeiy	Joerger and Haas (1994)
R. delemar	V209W >C4	binding of fatty acids tributyrin relative to triolein	2 fold increase of	Joerger and Haas (1994)
R. delemar	V209W F112W	binding of fatty acids >C4	80 fold increase of tributyrin relative to tricaprylin; no triolein hydrolysis	Klein *et al*. (1997)
R. oryzae	F214Y	binding of fatty acids >C12	20% increase of caproic acid methyl ester relative to oleic an stearic acid	Atomi *et al*. (1996)
R.oryzae	F216R	mutant should stabilize Δ9 double bond	90% increase of oleic acid methyl ester relative to stearic acid	Atomi *et al*. (1996)

not fully understood.

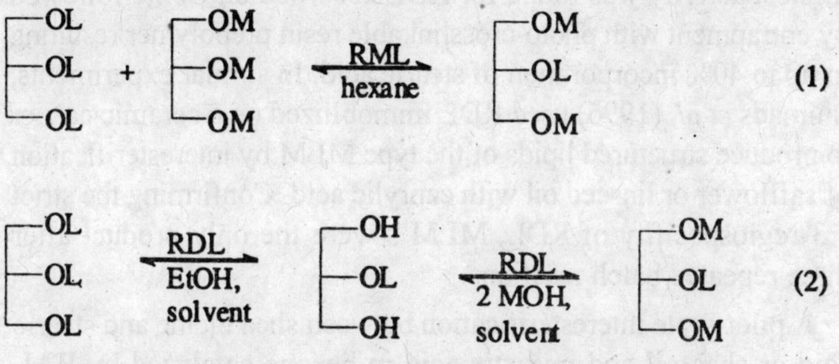

OL + OM → (RML, hexane) → OM / OL / OM (1)

OL / OL / OL → (RDL, EtOH, solvent) → OH / OL / OH → (RDL, 2 MOH, solvent) → OM / OL / OM (2)

Figure 9.5: ***Production of structured lipids by direct interesterification (1) or a combination of alcoholysis and esterification (2)***

APPLICATIONS

Lipid Modification

The natural substrates of lipases are triacylglycerols and, as a consequence, most applications of lipases deal with the modification of lipids. The low reaction enthalpy of lipase hydrolysis allows to use these enzymes for hydrolysis as well as for the synthesis or interesterification of triacylglycerols. Taking advantage of the mild reaction conditions (permitting the use of labile fatty acids as substrates), lipase regioselectivity, chain-length and fatty acid specificity, the production of structured lipids with a defined distribution of fatty acids along the glycerol backbone (e.g cocoa-butter equivalent, Betapol™) and the synthesis of emulsifiers like mono- or diglycerides (MAG or DAG) were extensively investigated.

As shown by Berger and Schneider (1991) and Waldinger and Schneider (1996), lipases from *Rhizopus sp.* have a strict 1,3-regioselectivity in nonpolar organic solvents where acyl migration is slow. The highly 1,3-selective RDL (Amano D) reacted at the primary position 76 times faster than at the secondary position, while the moderately selective CVL and RML reacted 26 and 11 times faster, respectively.

Intersterification between triglycerides

RDL was used to produce a cocoa-butter like fat by

interesterification between olive oil and stearic acid in hexane. The highest activity was found for RDL absorbed on Celite followed by entrapment with photo-crosslinkable resin prepolymer resulting in up to 40% incorporation of stearic acid. In similar experiments, Shimads *et al,* (1996) used RDL immobilized on a ceramic carrier to produce structured lipids of the type MLM by interesterification of safflower or linseed oil with caprylic acid. Confirming the strict 1.3-regiospecifity of RDL, MLM's were the only product after three repeated batch reactions.

A pilot-scale interesterification between shea oleine and stearic acid or shea oil and myristic acid in hexane catalyzed by RAL immobilized on Hyflo Super Cel was performed in a column reactor. Up to 100% conversion of fatty acid was observed and the enzyme lost almost no activity even after 166 h of operation. The interesterification of palm oil midfraction with stearic acid using Celite RAL in a lecithin-based micellar system in a gas-lift reactor resulted in a 2.8 fold higher productivity and reduced reaction times, if compared to shake flask experiments.

Butter fat was investigated as another suitable starting material for interesterification. Using RNL in a cosurfactant free microemulsion system made up from Span 60, Tween 60 or lecithin, a high proportion of C18:1 at the *sn*2 position was found. ROL immobilized on controlled pore glass was used in the esterification of milk fat with oleic acid in isooctane resulting in a lowered concentration of palmitic acid in milk fat and, consequently, in a lower transition temperature.

In order to increase the activity of the lipase in the presence of an organic solvent, the enzyme was coated with a lipid or surfactant. For example, a water-insoluble complex containing approximately 10 wt% of protein formed upon mixing aqueous solutions of lipases from RDL or RNL and a nonionic amphiphile (didodecyl *N*-D-glucono-L-glutamate. The modified lipase was soluble in most organic solvents and was > 100 times more active than suspended enzyme in the synthesis of di-and triglycerides from lauric acid and monolaurin in benzene. Similar observations were made by Goto *et al.* who coated a lipase from *Rhizopus sp.*

with the nonionic surfactant glutamic acid dioleyl ester ribitol amide and investigated the interesterification between lauric acid and benzyl alcohol in organic solvents.

RJL was coated with sorbitan monostearate and the lipid-coated enzyme was employed in a hollow fiber membrane reactor and the interesterification between tripalmitin and stearic acid in hexane was investigated. After reaching steady state (4 h), no loss in enzyme activity was observed within 76 h, and PPS and SPS were the major products.

Recently, structured lipids containing essential fatty acids, especially polyunsaturated fatty acids (PUFA) like docosahexaenoic or eicosapentaenoic acid, have been targets of lipase research. Fish oils contain up to 30% PUFA, mainly in the form of triacylglycerols, and are sensitive to elected temperatures, extreme pH etc. resulting in side reactions such as oxidation, *cis-trans* isomerization or double bond migrations. Yadwad *et al,* (1991) described the RNA- catalyzed glycerolysis of cod liver oil (9.6% PUFA) to yield 1 (3) monoacylglycerides with 29% PUFA. Shimada *et al.* (1996) interesterified tuna oil with caprylic acid with RDL, immobilized on a ceramic carrier, and found that after 2 d at 30°C approx. 65% of the fatty acids at 1- and 3-position in tuna oil were replaced by caprylic acid. Moreover, the enzyme could be reused 14 times without significant loss of activity.

Recently, we described the synthesis of highly pure triglycerides of the MLM type, which are important in nutrition (e.g. for patients with pancreatic insufficiency). Initially, we investigated the direct interesterification between a triglyceride bearing long-chain fatty acids (e.g. triolein) with a short-chain triglyceride (e.g tricaprylin) in *n*-hexane. RML was found most suitable, but the highest yield of MLM was only 31%. Significantly higher yields and purities were obtained by a combination of alcoholysis and esterification (Figure 5). In the first step, a triglyceride (e.g. triolein, peanut oil) is subjected to alcoholysis with ethanol at controlled water activity in methyl-*t*-butyl ether yielding 2-monoglycerides (2-MAG). RDL was found most suitable and after crystallization up to 72% yield of 2-MAG were achieved. The subsequent esterification of this 2-MAG with caprylic acid in an organic solvent catalyzed by RDL gave a highly pure MLM-triglyceride, which contained more than 90% caprylic acid in *sn*1- and *sn*3 position and 98.5% unsaturated fatty acids in *sn*2-position. Beside RDL, RJL, was also suitable in this reaction.

Rhizopus sp. lipase were also found active in supercritical carbon dioxide, as shown by Nakamura *et al*. (1986(for the RDL-catalyzed interesterification of triolein and stearic acid.

Preparation of MAG or DAG

Monoacylglycerides (MAG) as well as diacylglycerides (DAG) are widely used emulsifiers in the food and pharmaceutical area. Although numerous articles are dealing with the lipase-catalyzed synthesis of these compounds. Most MAG and DAG are still chemically produced. Examples for the lipase-catalyzed synthesis of MAG have recently been revised, and in several cases lipases from *Rhizopus species* have been used.

TABLE 9.5: SYNTHESIS OF DI- OR MONOACYLGLYCERIDES USING LIPASE FROM *RHIZOPUS SPECIES*

Acyl acceptor	Acyl donor	Reaction system	Yield (%)	Lipase	Reference
glycerol	plamitic acid	solid-phase	80 (1,3)[b]	RJL	Weiss (1990)
glycerol	vinyl caprylate	MTBE[a]	80 (1,3)[b]	RDL	Berger *et al*. (1992)
glycerol	ethyl caproate	free evaporation/ precipitate	80 (1,3)[b]	RAL	Millqvist-Fureby *et al*. (1996a)
water[1]	palm oil	reverse micelles	(26) 78[c,d]	RDL	Hǫlmberg and Österberg (1988)
water[1]	castor oil	buffer, pH 7.5	(8) 23[c]	RAL	Flenker and Spener (1990)
ethanol[1]	tripalmitin	MTBE	(32) 97[c,d]	RAL	Millqvist *et al*. (1994)
glycerol[2]	palm oil	reverse mcelles	(30) 30[c]	RAL	Holmberg *et al*. (1989)
glycerol[3]	lauric acid/	reverse micelles	(20) 62[c]	RDL	Hayes and Gulari (1994, 1992, oleic acid 1991)

water	crambe oil	reverse micelles	?	RJL	Derksen and Cuperus (1992)
glycerol[3]	oleic acid	water	(20) 60[c]	RDL	Tsujisaka *et al.* (1977)
glycerol[3]	oleic acid	org. solvent. $SCCO_2$	(6–15) 18–44[c]	RAL	Gancet (1990)
glycerol[3]	palmitic acid	hexane	(20) 61[c]	RDL	Kwon *et al.* (1995)

[1]: hydrolysis or alcoholysis of triacylglycerides to 2-MAG's, [2]: glycerolysis of triacylglycerides yielding mixtures of 1 (3)-MAG's and 2-MAG's, [3]:esterification or transesterification of glycerol with fatty acids or esters yielding 1 (3)-MAG's; [a]methyl *t*-butyl ether; [b]1, 3: 1,3-DAG; [c]overall yields of MAG relative to glycerolysis (per mol TAG), hydrolysis (per mol TAG) or esterification (per mol glycerol) are given in brackets; [d]2-MAG

DAG have been synthesized in high yields (up to 80%) by esterifications between glycerol and fatty acids (or their derivatives) using lipases from *Rhizopus delemar, arrhizus* or *javanicus* and 1,3-DAG was the major product. For example, Berger *et al.* (1992) absorbed glycerol onto silica gel to create a large surface area. RML-catalyzed esterification with fatty acid methyl esters or RDL- catalyzed esterification with fatty acid vinyl esters provided 1,3-DAG in good yields (>70%) and high regioisomeric purity (>98%). Vinyl ester reacted faster, but are also more expensive. Although RDL was more regioselective, RML was preferred because it was more active. Several unsaturated or hydroxy fatty acids were also converted to MAG or DAG in high yields with RNL using the same method.

MAG were prepared by lipase hydrolysis of oil in reverse micelles or buffer, but yields were not satisfactory, mainly due to rapid acyl migration from 2-MAG to 1-or 3-MAG, which is further hydrolyzed by the lipase. This problem was overcome by performing an alcoholysis of triglycerides at controlled water activity. Thus, the reaction between tripalmitin with ethanol using a lipase from *Rhizopus arrhizus* immobilized on Celite under optimized conditions gave 97% yield of 2-MAG. Alternatively, eaterification between glycerol and fatty acid in various reaction systems with *Rhizopus* lipase was studied by several groups. Hayes and Gulari (1991) employed a 1.3-specific lipase from *Rhizopus delemar,* but mainly 1 (3)-MAG was formed, which again was related to acyl migration. Conversion reached 50 to

60%, and the reaction rate decreased in the order oleic acid > caprylic acid > myristic acid = lauric acid > stearic acid = palmitic acid. A third general approach is based on the glycerolysis of fats or oils. Gancet (1990) developed a two fixed-bed segment for the continuous glycerolysis of beef tallow. At a residence time of 75 min, a yield of 38.8 g MAG per 100 G beef tallow was reported. The reactor system was run for several months with good operational stability of the mycelial lipase from *Rhizopus arrhizus*.

Triacylglycerol hydrolysis

A comparison of several lipases in the hydrolysis of crambe oil (rich in erucic acid) in reverse micelles revealed that most lipases followed Michaelis-Menten kinetics, and RJL was found to be the most active enzyme. A similar system was studied by Kim and Chung (1989) using RAL and palm kernel olein in AOT/ isooctane reversed micelles. In another study it was found that a cationic surfactant (CTAB) decreased the maximal rate of a RDL-catalyzed hydrolysis of triolein by a factor of 50, if compared to AOT.

Organic Synthesis

Most lipases exhibit high stability in organic solvents, and their ability to discriminate between stereoisomers has lead to a large number of applications for the resolution of a wide variety of choral alcohols or acids. For these investigations, mainly lipases from *Pseudomonas sp., Candida sp., Rhizomucor sp.* and from Porcine pancreas were used. Only a few examples deal with lipases from *Rhizopus sp.* — despite the fact that these lipases accept vinyl acetate as an irreversible acylating agent and are stable toward the acetaldehyde concomitantly formed.

Goto *et al.*(1996) studied the kinetic resolution of ibuprofen using surfactant-coated lipases. They found that surfactant-coated RJL catalyzed efficiently the acylation of (*S*)-ibuprofen with hexadecanol in isooctane, and up to 70%ee were achieved. Moreover, the coated lipase was approx. 100 times more active than the powdered enzyme.

In another example, RJL was used to prepare a key precursor

in the synthesis of tautomycin. RJL allowed for high conversion and reaction rates in the acylation of a *meso*-triol with vinyl acetate (53%ee, 1 d) or the hydrolysis of the corresponding triester (93%ee, 10 d)

Excellent optical purities (>95%ee) and good yields (64–95%) were reported in the RDL-catalyzed hydrolysis of *meso*-diacetates employed in the synthesis of the antiviral carbocyclic nucleosides (–) -carbovir and (–)-BCA. Here, RDL was superior to lipases from *Pseudomonas sp*. or Porcine pancreas.

RJL was used in the resolution of *cis*-3-acetyloxy-4-phenyl-2-azetidinone, a compound important in the semi-synthesis of taxol. Optical purities were excellent (98.5%ee), but yields were lower compared to other lipases.

Lipases can be used to deprotect carboxyl groups in peptides. Braun *et al*. (1991, 1992) cleaved the heptyl ester carboxyl protective groups from a wide range of dipeptides using RNL (Amano N). This lipase did not cleave the amino protective groups Cbz, Boc, Aloc, or Fmoc and the heptyl protective groups survived conditions used to remove these amino protective groups (hydrogenation, HCL/ether, or Pd(0)/C-nucleophile). Although the crude lipase (Amano N) also hydrolyzed the peptide link, pretreatment with PMSF, a serine protease inhibitor, eliminated this side reaction. Hydrolysis of the heptyl group slowed and sometimes did not proceed when the peptide was hindered and/or hydrophobic.

OUTLOOK

Lipases obtained from the genus *Mucorales* have been explored in a wide range of synthetic reactions in geochemistry and organic synthesis. *Rhizopus* lipases have found somewhat less applications, than the lipase from *Rhizomucor miehei* which is used, in immobilized form, in some geochemical biotransformations. However, *Mucorales* lipases can be heterlogously expressed in active form, without they need to apply cumbersome refolding procedures. In combination with the better understanding of the structure and function of these lipases, the way is open to taylor these stable and active enzymes to the needs of the organic chemist, and the applications given above already show that *Rhizopus* lipases are versatile catalysts. It is to be expected that *Rhizopus* lipases may be among those enzymes who will find increasing applications in industrial enzyme technology.

10

Microbial Lipases

Lipases, beside their use as drugs in disorders of the digestive system, have proved to be useful catalysts in biotechnology.

In the last years lipases of different sources have been recognised as valuable catalysts for organic transformations both on the laboratory and on the industrial scale. While bulk application already enjoyed a long standing tradition in areas of food processing and detergent formulations, new examples of the beneficial use of lipases have now emerged in the production of examples of the beneficial use of lipases have now emerged in the production of fine chemicals. In particular the stereoselectivity of lipases has been employed to manufacture enantiomerically pure pharmaceuticals, agrochemicals and food additives.

Their broad synthetic potential is largely due to the fact that lipases can accommodate a wide range of substrates other than triglycerides, such as, esters, thioeters and amides.

The stability in non-aqueous organic solvents, expanding the synthetic potential of lipases, can be applied to hydrolysis, ester synthesis and transesterifications.

LIPASES APPLICATIONS

Flavour and Feed Industry

The majority of flavours and fragrances used in foods and perfumery are esters, some of which are fatty acid esters. Many of the naturally occurring esters are one synthesized, and lipase biocatalysis could be advantageous due to specific and mild reactions from fatty acids and glycerol (branched, cyclic, aromatic, very short chain) lipases may still accept these. Synthesis of terpenes alcohol esters by various lipases has been reported, as well as the continuous synthesis of ethyl butyrate (pineapple/ banana flavour) by immobilised *Candida rugosa* lipase. A number of other flavour esters were made. Biochemical production of butter flavour can be made by alcoholysis of butter fat and ethanol, e.g. with *Candida rugosa* lipase. This type of reaction provides a mixture of fatty acid ethyl esters, which may be combined with natural components in foods, e.g. margarine and cookies. The oil industry is widely interested in the application of lipase to the neutralization of oil acidity. The acidity of tropical oils is due to the presence of free fatty acids. The content of these acids can be depressed by partial esterifications by lipases. The Soy-bean oil changes its flavour during storage and during the frying process giving rise to a typical fish smell. The Soy-bean oil flavour instability is due to the linoleic acid. At low temperature *Rhyzomucor miehei* lipase has shown specificity towards polyinsaturated fatty acid. It is possible to decrease the linoleic acid to 3% content by biocatalised transesterification at 10°C. Otherwise γ-linoleic acid has an high dietetic and commercial value. It is in the prostaglandin path-synthesis and it is rare in edible oils. At the moment the natural source of this fatty acid is evening primrose seeds. In order to obtain a fraction with an higher content of γ-linoleic acid, from evening primrose oil as starter, turnip lipase was employed. Using similar techniques it is possible to get hypocaloric products enriching oil with short and medium chain-length fatty acids.

The properties of fat depend on the fatty acids content and the commercial value of one fat compared to another is based on its fatty acid structure. Traditionally, fat and oil processors have changed the fatty acid structure of their materials by blending different triglyceride mixtures, by chemical modification of the fatty acids or by rearrangement of the fatty acids on the glyceride backbone of the fat (interesterification). Specific enzyme catalysts may be used for interesterification. In this field, another example of lipase application in the cookies industry is cocoa butter production. This lipid contains an high amount of stearic acid and is composed of 1-palmitoyl-2-oleyl-3-stearoyl glycerol. An alternative source of it at cheaper cost is palm oil. In fact palm oil is composed of 1.3-dipalmitoyl-2-oleyl glycerol. The 1,3 regioselective transesterification, catalyzed by *Rhyzomucor miehei* lipase, of palm oil yields cocoa butter. *Rhizopus delemar* lipase shows the same regioselectivity, it is able to biotransform olive oil into 1,3-distearyl-2-oleyl glycerol. In the same way the immobilized lipase from *Humicala lanuginosa*, a thermostable enzyme catalyzes the transesterification of olive oil into 1, 3 - distearyl - 2 oleyl glyceral at 50°C in 30 h with 65% yield in stearic acid.

Flavour development of cheeses is to a great extent due to the ability of lipases to modify milk fat by partial hydrolysis. Specific free fatty acid profiles are generated by naturally occurring microbial lipases or enzymes added for flavour enhancement. In the first case, moulds involved in the ripening of certain varieties of cheeses, e.g. *Penicillium sp*., produce lipases. *Penicillium roqueforti* lipases have a short-chain fatty acids specificity and the liberation of a suitable balance of mainly short-chain, volatile fatty acids seems to be the general effect behind a characteristic flavour. Probably some of the fatty acids may be released as ethyl- or other alcohol esters with flavour potential, by lipase catalysed alcoholysis or esterification. Pre-gastric esterases are used in the manufacture of Italian cheeses to produce the characteristic piquant flavour.

Flavour development is usually the result of a very small degree of lipid hydrolysis (few percent) whereas a higher degree may lead also to a change in the physical properties of the fat. Such partial hydrolysis of oils and fats can improve the palatability and digestibility of the lipid component in feeds and pet foods. Hard fats may be softened due to mono-and di-glycerides and more energy supplied to the animals if the lipid is 'pre-digested'.

The production of emulsifiers is another important aspect of the food industry. At the present lecithin, monoglycerides and sugar esterified with fatty acids are employed as emulsifiers in foods. Lecithin is a by-product of seed oil refining. Enzymes are used to produce lyso-lectihin which has better emulsifying properties than normal lecithin. Lyso-lecithin is used in margarine and cosmetics, for instance.

Detergent industry

Fatty acids and oil soiling are difficult to remove from fabrics at low washing temperatures. Spillage of lipid containing foods and accumulation of sebum from skin are common problems. Therefore, intensive research efforts have been directed towards development not only of new detergent compounds and surfactants but also toward enzymes which could facilitate removal of this type of soiling under wash conditions. Hydrolysis of triglycerides and other fatty esters does not increase the water solubility of the lipids as readily as in the case of protein and starch hydrolysis. It has been shown that a mixture with partial glycerides and free fatty acid is more effectively removed from the fabric than free fatty acids alone are, so that a non-specific lipase is not necessary or even preferable in washing. A detergent lipases require an high alkaline stability (common powder detergents pH is 9–10), are thermostable enzymes stable in the presence of protease and compatible with common detergents. The first detergent lipase on the market was Lipolase^{TM}, introduced early in 1988 by Novo. It was also the first industrial enzyme produced by recombinant technology.

Pulp and Paper

Enzyme treatments, including the hydrolytic application of lipases, have recently been taken up by the paper and pulp industry. Although lipids constitute a very small part of wood, they may have important effects. In spruce, about 2% resin is found. The resin is a mixture of triglycerides, free fatty acids and other components. The resin content is reduced to about 0.5% or less in existing pulping processes. However a more complete resin removal is possible by the addition of lipase in the process, with improved water absorption characteristics of cellulose fibers as the result. The resin content may also influence other properties, e.g. paper strength or colour and result in paper making processes (the formation of suspension of wood fibres). A lipase preparation, Resinase^{TM}A is produced by Novo and used in pulping process. Printed paper can be treated by lipase in order to facilitate removal of colour from the printing ink, containing vegetable oil base. This application can be useful for the improvement of return paper.

Chiral Switch

There is an increasing trend towards the use of optically pure stereoisomers with pharmaceutical activity, which are more target-specific and show fewer side-effects than racemic mixtures of isomers. This leads to an increasing demand for efficient processes for the industrial-scale synthesis of optically active compounds and to an increased research and production of chiral drugs via, for instance, lipases or esterases employment. The biocatalytic strategy to the resolution of racemic mixture can be applied to compounds containing alcohols, esters and acid moieties. If chiral or prochiral substrates are being used, it is usually just one enantiomer which undergoes reaction, thus leading to the chiral resolution of racemates.

LIPASE CATALYTIC ACTIVITY AND SELECTIVITY

In organic solvents, lipases catalyse the transfer of acyl groups from suitable donors to a wide array of acceptors other than water. Depending on the type of acyl donors and acceptors, lipase-catalysed reactions include esterifications, transesterification, amidations, peptide synthesis and macrocyclic lactone formations.

As well known, the quantification of enzyme selectivity can be determined using the so-called "Enantiomeric Ratio", or E-value, and it serves this purpose as it summarizes the enantioselectivity of an enzyme in a single number. The E-value has found its way both as a process parameter for the enzyme-catalysed kinetic resolutions, as well as for the evaluation of the intrinsic structure-function relations underlying enzymatic performance.

The presence of competing enzymes in the crude lipase

preparations should also be considered. In general, lipases from microbial sources are produced extracellularly and are virtually homogeneous in terms of lipolytic activity. In contrast, crude mammalian, fungal and plant lipases preparations can contain several other interfering enzymes, including proteases and esterases, which may possess opposite or poor streoselectivity as compared with the lipases. As a consequence, low optical yields or unreproducible results are often encountered in using these crude lipases, depending on the state of the enzyme preparations. Therefore, to enhance the optical yield, a number of precautions can be taken. These include treatment of the crude enzyme preparation with chemical reagents (e.g. serine protease inhibitors) to inactivate interfering enzymes; treatment of the crude enzyme preparation by physical means (e.g. partial protein purification; lyophilization) to remove or selectively inactivate competing enzymes; selection of a suitable acyl donor which serves as a poorer substrate for intefering enzymes and/or as a more efficient substrate for the target enzymes.

Although the great potentials of many lipases in stereoselective transformations, their performance under conditions prevailing in e.g. industrial reactors, in general, is far from optimization; in only a few cases this potential is reflected in an industrial patent ownership. It is therefore important to improve the properties, stability and selectivity, of lipases by using different strategies such as, for example site-directed mutagenesis or protein engineering, or trying to modulate enzyme selectivity and activity employing as reaction media non conventional solvents such as organic solvents or supercritical fluids.

Within the past ten year, lipase-encoding mutated genes from several organisms have been cloned. Some of these have been expressed in their new genetic background and recently X-ray diffraction studies have produced models of the three dimensional structure of several lipases. Also within this frame, biochemical and molecular genetic data have been gathered that allow the identification, in some lipases, of those amino acids that play a crucial role in the enzyme activity. These areas of research have converged to the extent that a co-ordinated approach. using computer-assisted molecular modelling and recombinant DNA technology, to examine structure-function relationships in these enzymes is now possible.

As far as the "solvent approach" has been concerned during the last decade, the interest for enzymatic reactions in organic media has been intense. Many fundamental studies have been carried out and a few industrial processes have been developed. The composition of the reaction medium is decisive with respect to the equilibrium position and substrate specificity. The degree of hydration of the medium and the biocatalyst is acknowledged to be of extreme importance. The thermodynamic water activity (aw) is used to describe the hydration level of these enzymatic systems. Minute amounts of water are necessary for the enzyme to maintain an active conformation. Any further increase of water, results in the enhancement of protein flexibility thus affecting enzyme stability. Recent studies have elucidated some effects of the hydration level on both activity and stability of enzymes in dehydrated environments. The interaction between a protein molecule and the surrounding water leads to a greater conformational mobility of enzymes. Consequently the hydration level is a critical parameter to take into account in order to optimise non-aqueous enzymatic processes.

As well known, lipases have prevent to be versatile and selective catalysts for the hydrolysis of a large range of interesting esters. In organic media, they catalyze the reverse reaction leading to various combinations of chiral alcohols and/or chiral acids.

The effect of solvent choice has been addressed as "medium engineering" since this might afford a relatively simple way for the fine tuning of enzyme activity, chemioselectivity, regioselectivity and enantioselectivity. The possibility to vary the activity and the selectivity of the lipase biocatalyzed reactions, by changing the nature of the reaction medium, can be considered a useful tool to modulate the chemical interactions between enzyme, substrate and reaction media with different physico-chemical properties.

Very recently, the use of microbial lipases in non-conventional solvents, such as supercritical fluids, has been proposed as a means of improving the activity and utility of such enzymes in anhydrous environments.

The potential use of supercritical fluids (SCFs) in chemical separation processes has been of considerable research interest for the past decade. The fundamentals of SCF extraction technology and a number of potential applications have been described in several review papers. One very interesting and, as yet, not fully tested offshoot of SCF extraction technology is the use of an SCF solvent as a reaction medium in which an SCF either actively participates in the reaction or functions only as the solvent medium for reactants or catalysts and/or products.

By exploiting the unique solvent properties of SCFs (wide variations in density and viscosity are possible with small changes in pressure and/or temperature; Table 10.1), it may be possible to enhance reaction rates while maintaining or improving selectivity. Also, separating products from reactants can be greatly facilitated by the ease with which the solvent power of the SCF can be adjusted.

TABLE 10.1: PHYSICAL PROPERTIES OF SUPERCRITICAL FLUIDS

	Gas	Supercritical fluid	Liquid
Density (g/cm^3)	10^{-3}	0.1–1	1
Diffusion (cm^2/s)	10^{-1}	10^{-3}–10^{-4}	$<10^{-5}$
Viscosity (g/cm*s)	10^{-4}	10^{-3}–10^{-4}	$<10^{-2}$

Gas Like:

High Diffusivity (mass transport in complex matrices)
Low Viscosity (favourable flow characteristics)

Liquid Like:

High Solvatating Power (dependent on density)

The present work reports some recent results and some general observations on parameters which can enhance the reactivity and the stereoselectivity in raw and recombinant lipase catalysed transesterifications of some model compounds, e.g. 1-phenylethanol, and some acyl donors, e.g. vinylacetate, in organic solvents and supercritical carbon dioxide.

ENZYME ACTIVITY

Before optimising the environment of an enzyme, we screened a number of wild type and recombinant lipases for the ability to catalyse the transesterification reactions between a secondary alcohol and vinylacetate, in some organic solvents (heptane, toluene and *tert*-buthylether) and supercritical carbon dioxide with the aim to describe the effect to the solvent on lipase activity.

We have chosen commercial lipases: four wild-type and recombinant from fungal source, and two recombinant from thermiphile microorganisms. These enzymes are usually employed as catalysts in organic synthesis and, in the case of RML even cheese manufactor. The 3D structures are available for only 3 of them (CRL, CAL A, CAL B, RML) in open and closed lid form. While CRL and RML show the "interfacial activation" phenomenon when employed in water systems for hydrolysis reactions, CAL B is active even with the substrate at lower concentration than CMC (Critical Micellar Concentration). The enzymes have different regioselectivity toward *sn*-position in triglyceride hydrolysis.

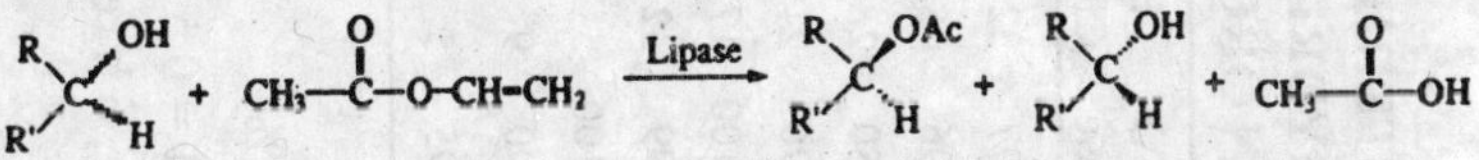

$$\mathrm{RR'C(H)OH} + \mathrm{CH_3{-}\overset{O}{\overset{\|}{C}}{-}O{-}CH{=}CH_2} \xrightarrow{\text{Lipase}} \mathrm{RR'C(H)OAc} + \mathrm{RR'C(H)OH} + \mathrm{CH_3{-}\overset{O}{\overset{\|}{C}}{-}OH}$$

Figure 10.1: ***General scheme of transesterification reactions of secondary alcohols and vinylacetate biocatalyzed by lipase***

The isoform composition of the employed enzymes was first checked by gel electrophoresis in non denaturing conditions. The results pointed out the presence of several isoforms with esterasic activity in wild-type enzyme preparations and in recombinant ones.

In our recent research work we studied esterification reactions of model substrates (phenylethanol derivatives) with acetic anhydride catalyzed by a bacterial lipase from *Pseudomonas cepacia*. The results pointed out the percentage of conversion and enatioselectivity dramatically depend on the utilized solvent, while

TABLE 10.2: TRANSESTERIFICATION REACTIONS OF (±) 1-PHENYLETHANOL WITH ACETIC ANHYDRIDE EMPLOYING IMMOBILISED *PSEUDOMONAS CEPACEA* LIPASE IN ORGANIC SOLVENTS AND IN SUPERCRITICAL CO_2

Substrate	SC-CO_2		Hexane		CCl_4		Toluene		Benzene		$CHCl_3$		NEt_3			
	C^a	e.e^b	C^a	e.e^b	C^a	e.e^b	C^a	e.e^b	C^a	e.e^b	C^a	e.e^b	C^a	e.e^b	C^a	e.e^b
1-phenyle-ethanol	46	97	32	42	28	88	28	88	21	87	36	93	20	67	14	85
1-(4-fluorophenyle)-ethanol	52	96	48	69	34	76	42	80	28	82	28	85	25	7	16	75
1-(4-chlorophenyl)-ethanol	52	80	28	58	27	69	23	70	18	67	28	90	24	12	18	85
1-(4-bromophenyl)-ethanol	55	96	21	88	25	70	24	89	20	73	27	80	18	4	11	84
1-(4-methoxyphenyl)-ethanol	29	100	17	60	28	49	26	88	25	86	29	55	19	20	29	39
1-(4-f*tert*-buthylphenyl)-ethanol	52	99	32	93	28	62	24	70	16	75	24	80	20	3	10	78
1-(2-bromophenly)-ethanol	2	100	5	71	3	13	2	100	2	100	1	100	10	4	1	31
1-(2-4-dichlorophenyl)-ethanol	10	90	9	76	4	75	8	75	7	60	2	60	14	88	2	43
1-(2-methoxyphenyl)-ethanol	7	95	5	10	5	13	1	54	2	14	3	7	15	1	3	16

[a] Conversion (%)

[b]Enantiomeric excess of product (%)

Reaction condition: Phenylethanol = 0.5 mM, acetic anhydride = 0.5 mM, reaction volume = 1.75 mL, T = 40°C, enzyme preparation = 10 mg, 6 h reaction time

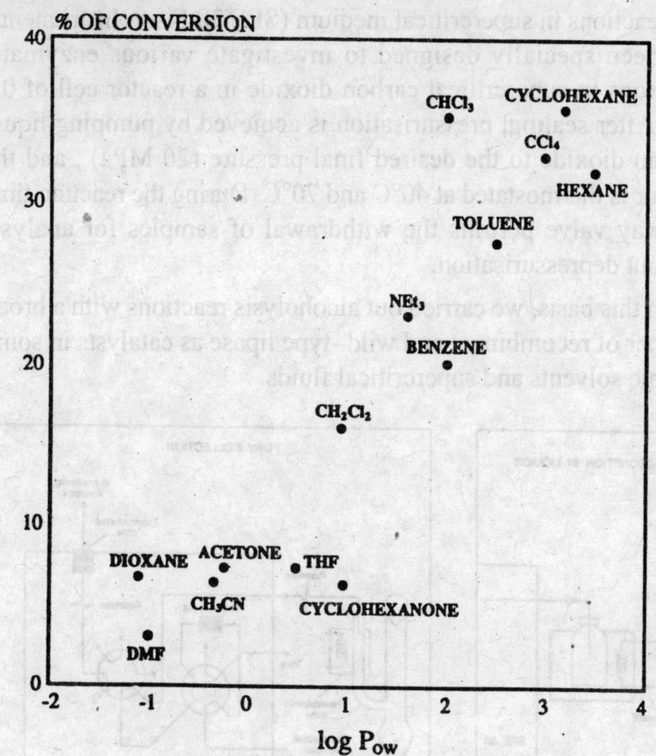

Figure 10.2: ***Esterification reactions of 1-phenylethanol and acetic anhydride catalyzed by*** **Pseudomonas cepacia** ***lipase in organic solvents. Conversion (%) versus solvent characteristics expressed as logPow. (DMF: dimethylformamide, THF: tetrahydrofuran, NEt_3: triethylamine, CCl_4: carbon tetrachloride, CH_2Cl_2: dichlorometance, $CHCl_3$: chloroform, CH_3CN, acetonitrile)***

the use of differently activated substrates did not substantially improve the reaction rates in the employed conditions,. Products were analysed by gas chromatography (Fisons 5300) using two serial capillary columns: OV1: 25m × 0.32 mm ID and DMePeBCDX, mega 25m × 0.25 mm ID. Data proved that there is a correlation between solvent hydrophobicity, expressed in terms of logPow values and percentage of conversion,. Moreover the results obtained in supercritical fluids showed an higher extent of conversion and enantiomeric excess values compared to those obtained in organic solvents. The apparatus employed to carry out the reactions in supercritical medium (SFE 30 Fison Instruments) has been specially designed to investigate various enzymatic reactions in supercritical carbon dioxide in a reactor cell of 0.9 mL. After sealing, pressurisation is achieved by pumping liquid carbon dioxide to the desired final pressure (20 MPa) , and the reactor is thermostated at 40°C and 70°C. During the reaction time a 6-way valve permits the withdrawal of samples for analysis without depressurisation.

On this basis, we carried out alcoholysis reactions with a broad number of recombinant and wild -type lipase as catalysts in some organic solvents and supercritical fluids.

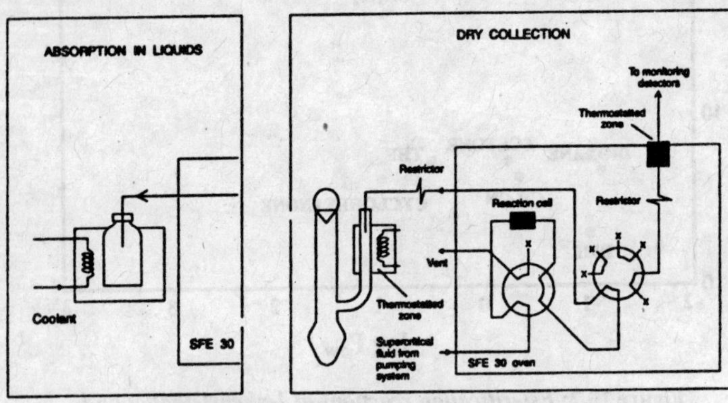

Figure 10.3: ***Schematic diagram of supercritical fluid bioreactor***

As expected, the use of hydrophylic organic solvents (*tert*-butylether and toluence) decrease the percent of conversion in comparison with the hydrophobic one (heptane). In some cases (wild-type and recombinant RML),, there is no detectable reaction product, showing remarkably different suitability of the substrate for the biocatalysts. As far as enzyme selectivity, the reaction product was proved to be the R (+) 1-phenylethyl acetate in all the reaction medium employed (e.e. 100%) unless in the case of recombinant CAL A, carried out in heptane, in which an opposite stereopreference was achieved, Table 4. In fact as far as the recombinant CAL A the reaction medium is able to modulate not only the extent of conversion but even the enzyme stereoselectivity.

TABLE 10.3: TRANSESTERIFICATION REACTIONS OF (±) 1-PHENYLETHANOL WITH VINYL ACETATE EMPLOYNG WILD-TYPE AND RECOMBINANT *RHYZOMUCOR MIEHEI* LIPASES

Enzyme	Solvent	Conversion (%)	Stereochemistry	e.e.$_P$ (%)
RML wt	$SCCO_2$	0.8	R(+) acetic ester	100
RML r	$SCCO_2$	1.27	R(+) acetic ester	100
RML wt	heptane	0	/	/
RML r	heptane	7,5	R(+) acetic ester	100
RML wt	toluence	0	/	/
RML r	toluence	0	/	/
RML wt	*tert*-buthylether	0	/	/
RML r	*tert*-buthylether	0	/	/

Reaction conditions: Phenylethanol = 1mM, vinylacetate = 10mM, reaction volume = 1mL, T = 40°C, enzyme = 5 mg, 24 h reaction time

TABLE 10.4: TRANSESTERIFICATION REACTIONS OF (±) 1-PHENYLETHANOL WITH VINYL ACETATE EMPLOYNG RECOMBINANT *ASPERGILLUS ORYZAE, CANDIDA ANTARCTICA* FORM A AND B LIPASES AND WILD-TYPE *CANDIDA RUGOSA* LIPASE

Enzyme	Solvent	Conversion (%)	Stereochemistry	e.e.$_P$ (%)
AOL r	$SCCO_2$	1.39	R(+) acetic ester	100
AOL r	heptane	6	”	
AOL r	toulence	/	/	/
AOL r	*tert*-butylether	/	/	/
AOL A r	$SCCO_2$	32.2	R(+) acetic ester	100
AOL A r	heptane	75.3	S(–) acetic ester	16
AOL A r	toluene	0.32	R(+) acetic ester	/
AOL A r	*tert*-bythlether	5,8	R(+) acetic ester	9,3
AOL B r	$SCCO_2$	/	/	/

AOL B r	heptane	10	R(+) acetic ester	.100
AOL B r	toluene	12,4	R(+) acetic ester	100
AOL B r	*tert*-buthylether	/	/	/
AOL wt	eptano	19,2	R(+) acetic ester	40
AOL wt	toluene	/	/	/
AOL wt	*tert*-buthylether	/	/	/

Reaction conditions: Phenylethanol = 1mM, vinylacetate = 10mM, reaction volume = 1mL, T°= 40°C, enzyme = 5 mg, 24 h reaction time

The use of recombinant thermostable lipase derived from thermophiles are likely to be more thermally robust and more stable in organic solvents than their mesophilic counter parts currently used in organic synthesis. In our experiments the recombinant thermostable lipase ESL-001-01 and ESL-001-02, supplied in partially purified form (60–95% pure) by Recombinant Biocatalysis Inc. showed in heptane a conversion value, after 24 hours reaction time, from 5 to 40% about, but a very poor selectivity. In *tert*-buthylether very low conversion product was achieved, Table 10.5.

On this basis the relationship between enzyme activity and solvent physical properties cannot be a simple correlation, and much will be learned from future experiments.

Carbon dioxide is the most commonly used supercritical fluid, and as such, we started our investigations of the lipase-catalyzed reactions in supercritical fluids in this solvent. Given the results in Table 4 and 5 we can observe that only with recombinant thermostable ESL-001-01 and ESL-001-02 lipases the percent of reaction product and the selectivity of biocatalytic reactions in supercritical fluids is higher than in conventional organic solvents. These are the first data reported for biocatalyzed reactions in $SCCO_2$ employing recombinant lipases.

As far the results obtained, the solvent hydrophobicity (carbon dioxide is almost non polar as hexane), the unusually high density ($0.754 g/cm_3$) and the high diffusivity of supercritical fluid (enzymes suspended in these media are subject to less internal diffusion limitations than those in conventional solvents) are probably the main reasons for the improvement of ester production.

TABLE 10.5: TRANSESTERIFICATION REACTION OF (±) 1-PHENYLETHANOL WITH VINYL ACETATE EMPLOYING RECOMBINANT THERMOPHILIC LIPASES

Enzyme	Solvent	Conversion (%)	Stereochemistry	e.e.$_P$ (%)
ESL-001-02	$SCCO_2$	12.26	R(+) acetic ester	100
ESL-001-02	heptane	5,4	R(+) acetic ester	68
ESL-001-02	toluene	0	R(+) acetic ester	/
ESL-001-01	$SCCO_2$	38,6	R(+) acetic ester	67,5
ESL-001-01	heptane	40	R(+) acetic ester	65
ESL-001-01	toluene	9.4	R(+) acetic ester	65

Reaction conditions: Phenylethanol = 1mM, vinylacetate = 10mM, reaction volume = 1mL, T = 70°C, enzyme = 1mg, 24 h reaction time

CONCLUSIONS

The use of enzymes in organic solvents is now commonplace, and analysis of the dependence of enzyme activity as a function of solvent will enhance our understanding of protein-environment interactions.

Supercritical fluids can be considered just as another nonaqueous media in which to attempt biocatalytic reactions. Since the physical properties of supercritical fluids are determined by temperature and pressure, these dispersants offer a very real opportunity to investigate the effect of solvent physical properties on an enzyme reaction in a controlled manner.

The work presented in this chapter was performed as first approach to investigate on supercritical fluid and conventional organic solvents suitability as reaction medium in transesterification reactions employing recombinant lipases of different sources.

Although the experimental data do not supply a predictive model, because solvent-enzyme interactions can be different according to the different substrates used, this approach could give rise to a side range of industrial applications through a careful optimisation of the reaction conditions and screening of more suitable enzymes. Moreover all recombinant enzymes used show an higher activity than wild-type ones. These data could be useful to protein engineering in order to optimise the enzyme performance through side-direct mutagenesis.

11

Industrial Enzymes by Protein Engineering

One of the major problems encountered in enzyme technology is the tendency of enzymes to get inactivated when exposed to non-physiological conditions. In particular, the fact that generally enzymes denature rapidly at temperatures above 50°C prevents them from being used in many industrial applications.

In the attempt to solve the problem of thermal inactivation of enzymes the industrial research is following two lines of activity. Firstly the search for thermostable enzymes from both thermophilic and hyperthermophilic bacteria has become a very popular and fruitful approach. Indeed, extremely robust enzymes able to survive at temperatures well above 70°C have already found their way to industrial processes. The second approach consists in the application of powerful protein engineering techniques which allow selective substitutions of amino acid residues that are crucial for thermostability. This approach, which is widely utilized nowadays, is well applicable when the 3-D structure of the protein to be stabilized is known. However, although the number of proteins studied at structural level is rapidly increasing, the structures of many industrially useful enzymes remain to be elucidated.

Can enzymes be stabilized by using novel molecular biology and protein engineering approaches even when their 3-D structure is unknown?

In the attempt to answer this question, thee industrial enzymes, the *Bacillus subtilis* neutral protease, the *Pseudomonas* isoamylase and carbamylase from *Agrobacterium radiobacter* were taken as model systems and were subjected to protein engineering following two different experimental strategies. From our data, as well as from the data of other laboratories, it appears that protein engineering can indeed become the strategy of choice whenever the stabilization of a protein is required.

BACILLUS SUBTILIS NEUTRAL PROTEASE

The *Bacillus subtilis* neutral protease is a zinc metalloendopeptidase extensively studied at both biochemical and genetic levels. Although its structure is not known, the enzyme belongs to a highly homologous family of proteases, three of which, the enzymes from *B. thermoproteolyticus, B. cereus* and *Pseudomonas aeruginosa*, have been crystallized and structurally characterized. The knowledge of the structures of these three enzymes prompted us to use computer-assisted techniques to model a 3-D structure of B, subtilis neutral protease. The model was then carefully scanned to localize possible residues which, if properly replaced, might contribute to enzyme stabilization. The rationale behind the selection of the residues to be replaced and the nature of the replacing amino acids is based on the well-known criteria of protein stabilization listed in. According to such criteria, any protein can be stabilized whenever site-directed mutations are created which 1) improve the hydrophobic interactions, 2) improve salt bridges, 3) stabilize α-helices by dipole capping, 4) decrease the entropy of unfolding and 5) create metal ion binding sites.

TABLE 11.1: RATIONAL APPROACH TO PROTEIN STABILIZATION

→	Improvement of hydrophobic interactions
	Improvement of salt bridges
→	α-helix dipole capping
	Improvement of hydrogen bonding
→	Decrease of entropy of unfolding
→	Addition of metal ion coordination sites

Our experimental approach was to produce a series of mutant enzymes, each carrying a mutation designed to satisfy one of the criteria listed in Table 1 (those indicated by the arrows). Therefore, a multiple mutant was obtained in which all the stabilizing mutations were added one after the other.

We started our stabilizing work from a neutral protease mutant selected in our laboratories which presents a Phenylalanine in a region (position 63) exposed to the solvent. We hypothesized that the replacement of the surrounding residues Ser9, Thr11, Thr17 and Thr61 with amino acids whose side chains could create a hydrophobic pocket around the aromatic ring of Phe63 would have stabilized the enzyme. The replacing amino acids were Val for Ser9, Arg for Thr11 and Gln for Thr17 and Thr61. The computer analysis predicted that the methylene groups of the side chains of the new residues would establish favourable hydrophobic interactions with Phe63. An additional mutation (Arg68>Leu) was added to remove the destabilizing electrostatic interaction between the guanidinium groups of Arg68 and Arg11. The stabilizing effect of the hydrophobic pocket modelled around Phe63 is given in Table 11.2 which shows that the T50 (the temperature at which 50% of the activity is lost after 30-minute incubation) of the (Phe63) neutral protease is enhanced by 3°C after the replacement of the five amino acids.

TABLE 11.2: ANALYSIS OF THE NEUTRAL PROTEASE MUTANTS

Mechanism	Mutant	ΔT50
Hydrophobic interactions	"Hydrophobic pocket"	+ 3.0°C
α-helix capping	K305D	+ 0.4°C
Entropic effect	A4T	+ 0.6°C
	G147A	+ 2.0°C
	G189A	+ 1.0°C
Metal ion coordination	"Ca loop"	+ 3.0°C
	MULTIPLE MUTANT	+ 9.9°c

It has been demonstrated that the interactions between the

extremities of the α-helix dipole and point charges favourably contribute to the enthalpy of the protein folded state. After scanning the seven α-helices of the neutral protease we found that the most C-terminal one, extending from residue 305 to residue 315, has a destabilizing positive charge at its N-terminus (Lys305). The replacement of Lys305 with Asp increase the T50 of the wild type enzyme by about 0.5°C.

One of the mechanisms by which proteins can be stabilized is by decreasing their entropy of unfolding. According to the thermodynamic principles, a folded protein (F) finds itself in equilibrium with its unfolded state (U).

U <---------> F

The thermodynamic equation which governs such equilibrium is:

$$\Delta G = G_F - G_U \quad (1)$$

or

$$\Delta G = (H_F - H_U) - T\,(S_F - S_U) \quad (2)$$

Equation (2) shows that any mutation able to decrease the entropy of the unfolded state (S_U) makes the ΔG value more negative and therefore stabilizes the protein in its folded conformation.

One way to decrease the entropy of unfolding is to replace amino acids having flexible backbones (e.g. glycine) with more rigid residues, provided that the substitutions do not create an unfavourable strain within the molecule.

With this in mind we selected three substitutions, Ala4> Thr, Gly147>Ala and Gly189>Ala, which turned out to increase the T50 of the wild type neutral protease by 0.5, 2 and 1°C, respectively.

In the *B. subtilis* neutral protease, the residues from 188 to 194 constitute a seven amino acid loop located in proximity of the catalytic site. In thermolysin, the highly stable enzyme belonging to the same protease family, the corresponding region is occupied by a 10-amino acid loop which defines a calcium coordination site.

After the replacement of the native 7-amino acid loop by the thermolysin loop, we have been able to prove that the mutant neutral protease acquires the ability to bind a third calcium ion (two calcium atoms are naturally bound to the enzyme and turn out to be essential for enzyme stability). The binding energy enhances the T50 value of the mutant enzyme with respect to the wild type by about 3°C, when the protein is exposed to calcium containing buffers.

Figure 11.1: ***3-D model structure of the multiple mutant of* B. subtilis *neutral protease obtained by side-directed mufagenesis. Indicated are the zinc atom essential for catalytic activity (fuchsia), the two calcium atoms required for stability (violet) and the third calcium atom (blue) bound to the mutant molecule after loop replacement (shown in yellow). The side chains of all amino acids which have been replaced are also shown in yellow colour***

Table 11.2 also shows the stabilizing effect of all mutations described above accumulated in a single mutant (Figure 11.1). Interestingly, a cumulative effect can be observed which results in a net increase in thermostability of the multiple mutant with respect to the (Phe63) neutral protease by about 10°C.

In conclusion, the data described above indicate that 1) a reliable 3-D protein structure can be predicted by computer-assisted molecular modeling whenever a homologous protein, the structure of which is known, does exist. 2). The computer derived model can be used to select specific amino acid substitutions able to enhance the protein stability on the basis of well defined general rules. 3) Once single stabilizing substitutions have been identified, multiple mutants can be created in which a cumulative stabilizing effect is generally obtained.

STABILIZING *PSEUDOMONAS* ISOAMYLASE

Isoamylases, enzyme which hydrolyse the α-1,6 glucosidic inter-chain linkages present in starch, are used industrially for the production of amylase and maltose. In the industrial production of maltose, the starch solution is first partially hydrolysed at high temperatures and thereafter the hydrolysis is brought to near completion by adding thermostable amylases. Finally, 97–98% conversion of starch into maltose is achieved by further addition of pullulanase and/or isoamylase.

Among the different isoamylases isolated so far, one of the best characterized both at biochemical and genetic levels is the isoamylase from *Pseudomonas amyloderamosa*. This enzyme is currently used by the starch industry but it has the disadvantage of being thermolabile and also of having a optimum pH for stability around 4. This implies that for the enzyme to be used in the industrial process the temperature of the reactor has to be cooled down considerably and the pH of the solution has to be adjusted (in the maltose production process amylases operate at pH 5.5–6). Therefore, for the improvement of the process, the availability of a thermostable isoamylase with an optimum of pH for stability around 6 would be particularly useful.

Since neither the structure of the enzyme is known nor a homologous enzyme the structure of which has been resolved is available, in order to isolate a pH dependent thermostable isoamylase mutant we adopted a random-site-directed mutagenesis strategy. Briefly, the experimental approach consists of the following steps: 1) PCR amplification of the isoamylase gene under conditions in which the polymerase makes an average of 1–2 errors per 100–200 nucleotide elongation, 2) preparation of a "library" of recombinant *B. subtilis* clones, each expressing a different isoamylase mutant, 3) screening of the library to identify the clones that produce isoamylase mutants with higher thermostability at pH 6. The library screening was carried out growing the recombinant clones on plates buffered at pH 6 and exposing the colonies to iodine vapours after their incubation for sixteen hours at temperatures between 37 to 45°C. Only the colonies producing thermostable isoamylase mutants were surrounded by blue halos when incubated at temperatures higher than 37°C.

To facilitate the subsequent characterization of the mutants, the isoamylase gene was first mutagenized to create unique restriction sites without changing the protein sequence, allowing the splitting of the gene into 5 restriction fragments. Therefore the PCR-mediated mutation was carried out on each fragment so as to restrict the region to be sequenced for identifying the stabilizing mutation (Figure 11.2). Table 11.3 summarizes the results of the screening procedure. One mutant of fragment 2 out of 85, 000 analyzed carried a thermostable mutation at position 355 (Thr>Scr) whereas 30 independent fragment 4 mutants were selected. The latter carried, together with different silent mutations, the stabilizing substitutions of Ile367 with either Thr or Ser. Asp507>Val was the relevant mutation present in the 9 mutants of fragment 4 identified by the screening of 90,000 clones. Finally, no thermostable mutant was selected out of 300,000 fragment 1 mutants and out of 12,000 fragment 5 mutants analyzed. To verify whether or not the stabilizing effect of each mutant was additive, double and triple mutants were obtained by site-directed mutagenesis and the thermostability of the purified enzymes was compared with respect to the wild type enzyme. As shown in the triple mutant Ile367>Thr/Asp507>Val/Thr355>Ser turned out to be the most stable enzyme, the double mutants

Ile367>Thr/Asp507>Val and Thr355>Ser/Ile367>Thr being more stable than the wild type but less stable than the triple mutant. Interestingly, when the stability assay was carried out at pH 4, no difference in thermostability between the mutants and the wild type was observed.

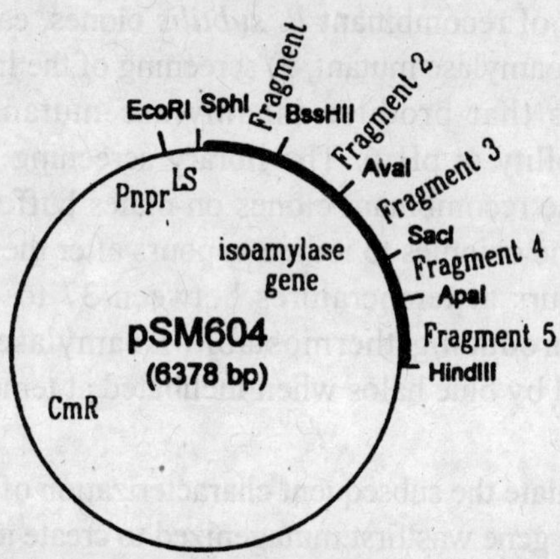

Figure 11.2 : ***Functional and restriction map of plasmid pSM604. Pnpr: promoter of the B, subtilis neutral protease gene under which the isoamylase gene is expressed. LS: leader sequence for secretion of subtilisin. The fragments of the isoamylase gene which have been mutagenized by PCR are indicated. For details see text.***

In conclusion, the random site-directed mutagenesis approach appears to be a powerful general strategy to isolate stability and/or activity mutants of any enzyme, provided that a rapid and reliable screening procedure is available. As in the case of neutral protease, the isoamylase stabilizing mutations turned out to exert a cumulative effect, the double mutants being more stable than both the wild type enzyme and the single mutants but less stable than the triple mutant.

TABLE 11.3: SCREENING OF THE ISOAMYLASE MUTANTS

Fragment	Number of colonies screened	Number of positive clones	Mutation
1	300,00	0	–
2	85,00	1	Thr355Ser

3	60,00	30	Ile367Thr
			Ile367Ser
4	90.00	9	Asp507Val
5	12,00	0	–

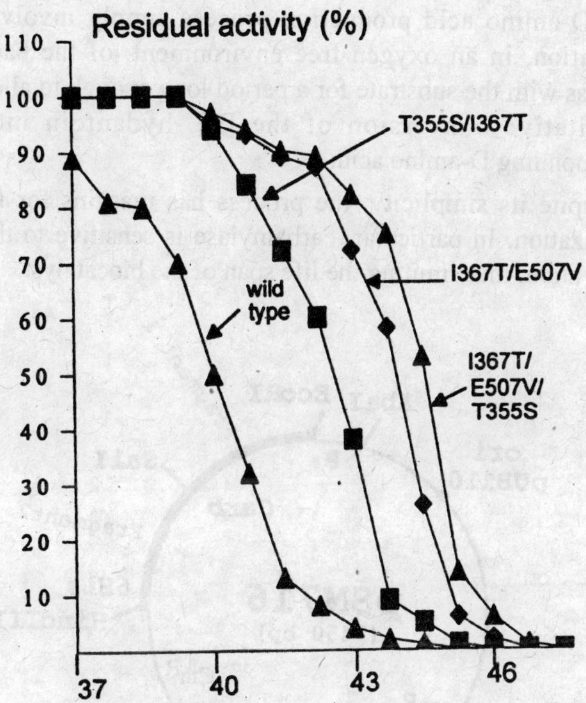

Figure 11.3: ***Thermal stability of isoamylase mutants. Residual activity was measured after 3-hour incubation at different temperatures***

STABILIZING CARBAMYLASE FROM *AGROBACTERIUM RADIOBACTER*

The D-amino acids D-phenylaglycine and D-p-hydroxyphenylglycine are among the most important optically active compounds for their use in the production of semisynthetic penicillins and cephalosporins.

These amino acids are currently synthesized in a two-step reaction process using as substrates D,L-5 monosubstituted hydantoins. According to this two-step process, the substrate is first hydrolysed to the D-carbamyl derivative by a D-specific hydantoinase and then further converted into the corresponding D-amino acid by an N-carbamyl-D-amino acid amidohydrolase (hereinafter carbamylase). Since several micro-organisms expressing both hydantionase and carbamylase have been isolated , the D-amino acid production process simply involves the incubation, in an oxygen-free environment, of the bacterial biomass with the substrate for a period long enough to allow the quantitative conversion of the D,L hydantoin into the corresponding D-amino acid.

Despite its simplicity, the process has margins for further optimization. In particular, carbamylase is sensitive to thermal inactivation, thus limiting the life span of the biocatalyst.

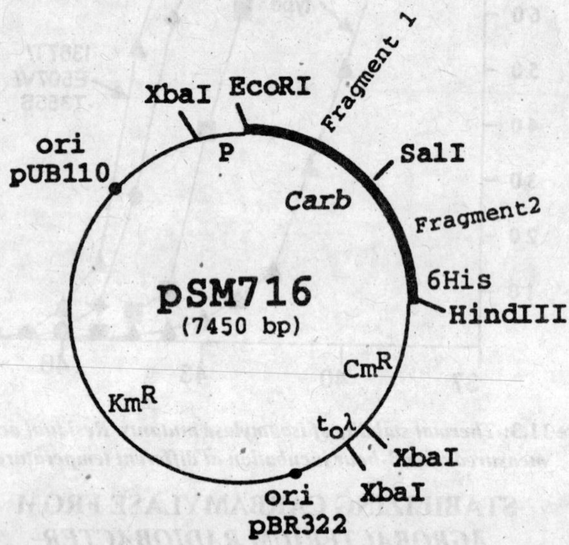

Figure 11.4 : ***Functional and restriction map of plasmid pSM716. P: costitutive promoter under which the carbamylase gene is expressed. 6 His: histidine tail add to the carboxy-terminal of the protein. The fragments of the carbamylase gene mutagenized by PCR are indicated. For details see Text***

In the attempt to select thermostable carbamylase mutants, we developed a random site-directed mutagenesis approach similar to the one described for the isoamylase.

The carbamylase gene was divided into two fragments and each fragment was amplified by PCR under conditions which favoured approximately 1–2 errors per 100 base pairs. Each amplified reaction was then ligated to plasmid pSM716 (Figure 11.4) and the two ligation mixtures used to transform *E. coli* cells. Two libraries were generated, each carrying random mutations at the 5' portion and at the 3' portion of the carbamylase gene, respectively.

The two libraries were replicated on nitrocellulose filters and after lysis of the colonies by freeze-thawing. the filters were incubated at 48°C for 16 hours, conditions that inactivate the wild type carbamylase. When the filters were laid on a medium containing the substrate N-carbamyl-D-phenlylglycine and the pH indicator Phenol Red, violet-red halos appeared around a few colonies, suggesting the presence of thermostable carbamylase mutants which, having survived the temperature treatment, converted the substrate to D-phenyl glycine and ammonia (Figure 11.5).

Two colonies were selected for further study. The sequence analysis of the carbamylase gene from these colonies revealed that a single point mutation had occurred in each mutant, both leading to the replacement of thereonine with alanine one at position 212 and the other at position 262.

TABLE 11.4: THERMAL STABILITY OF CARBAMYLASE MUTANTS

Mutation	ΔT50
Thr212Ala	3.3
Thr262Ala	2.3
Thr121Ala-Thr262-Ala	7.1

The carbamylase mutants were purified from *E. coli* cells as already described and the ΔT_{50} values determined using the purified proteins (T_{50} is the temperature at which the enzyme loses 50% of its activity after 30-minute incubation and ΔT_{50} is the difference between the T_{50} value of the mutant and the T_{50} value of wild type enzyme). As shown in Table 11.4, the Thr212>Ala mutant was 3.3°c more stable than the wild type carbamylase whereas Thr 262>Ala mutant had a ΔT_{50} value of 2.3. When the ΔT50 value was measured on the Thr212>Ala/Thr262>Ala double mutant, an increased thermal stability of 7.1°C was found indicating that, as in the case of neutral protease and isoamylase, stabilizing mutations have a cumulative effect when present on the same molecule.

12

Recombinant Gastric Lipases

For a long time, the occurrence of gastric lipolysis was attributed to contamination to the gastric content by pancreatic lipase as the result of duodeno-gastric reflux. Volhard (1901) established, however, that there exists a gastric enzyme capable of hydrolysing triacylglycerols in humans. In 1917, Hull and Keaton (1917) observed the presence of a dog gastric lipase under conditions precluding contamination by pancreatic lipase. Nevertheless, the real importance of gastric lipolysis was first understood very belatedly in studies on patients with cystic fibrosis as well as on chronic pancreatitis, where the amount of fat absorption amounted to 70% of the total ingested lipids, whereas the pancreatic lipase and bile salt levels were particularly low. In other patients, where pancreatic lipase is completely lacking due to a congenital deficiency, 50% of the ingested dietary fat was absorbed.

One particularity of all preduodenal lipases is that they are stable and hydrolyse their substrates under acidic conditions. Acidic lipases have been screened in various species (rat, mouse, calf, sheep, humans, rabbit, dog, cat, monkey, horse, hog, guinea pig) from the tongue to the pylorus. Up to now, six preduodenal lipases have been purified and characterized biochemically: rat lingual lipase, human gastric lipase (HGL) (Tiruppathi and

Balasubramanian, 1982), calf pharyngeal lipase, lamb pharyngeal lipase, dog gastric lipase, rabbit gastric lipase. These enzymes form a new family of lipases: the acidic lipases. The present review focuses on two enzymes belonging to this family: HGL and DGL. Their biochemical properties have by now been clearly established and HGL has been expressed *in vitro* in active form, using the baculovirus/insect cells system.

PHYSIOLOGICAL ASPECTS

Gastric lipases have been purified either from gastric juice or from fundic gastric mucosae and characterized *in vitro*. HGL was first purified by Tiruppathi and Balasubramanian (1982) and its secretion was stimulated by injecting pentagastrin. Nowadays, the enzyme is purified using cation-exchange (Mono S) chromatography followed by an immunoaffinity column procedure using monoclonal antibodies. Lipase activity can be measured potentiometrically on short-chain (tributyroylglycerol, TC4), medium-chain (trioctanoyl-glycerol, TC8) and long-chain (Intralipid™) triacylglycerols at 37°C using a pH-stat. The maximum specific activities are 1160 units/mg at pH 6, 1110 units/mg at pH 6 and 600 units/mg at pH 5, on TC4, TC8 and Intralipid™, respectively.

DGL was purified from the gastric juice of dogs with chronic gastric fistulae. Little if any DGL activity was observed in basal or stimulated gastric secretion, whereas the gastric mucosa was found to contain substantial amounts of lipolytic activity. This paradox can be explained by the occurrence of an irreversible inactivation of DGL at low pH levels, which are present in the stomach under fasting conditions. By buffering the acidic secretion *in vivo*, or by using an anti-acid secretion drug such as omeprazol during the stimulation, the effects of several gastric scretagogues on the secretion of DGL were determined. Unlike HGL secretion, DGL secretion was poorly stimulated by pentagastrin, while urecholine, 16,16-dimethyl prostaglandin E_2 and secretin were potent secretagogues. DGL was extracted from soaked dog gastric mucosa and purified after cation exchange, anion exchange and gel filtration chromatographies. The maximum specific activities were 550 units/mg on tributyroylglycerol at pH 5.5, 750 units/mg on trioctanoylglycerol at pH 6.5 and 950 units/mg on Intralipid™ at pH 4. DGL is nearly twice as active on long chain triacylglycerol as on short chain triacylglycerol.

HGL and DGL cDNAs were synthesized by RT-PCR amplification using stomach biopsies and reverse transcriptase was performed for 60 min at 37°C. The HGL and DGL coding regions 1.2 kb pairs and 1.55 kb pairs, respectively, were amplified by performing polymerase chain reaction (PCR) procedures. The cDNA coding for HGL and DGL were ligated in a pUC 18 plasmid. The deduced amino acid sequences of HGL and DGL were those of a 379 amino acid polypeptide, and the homology between the two proteins was about 85%.

The molecular masses of the HGL and DGL and 50 kDa. These enzymes are glycoproteins, containing around 15% carbohydrate. Endoglycosidase F treatment of HGL results in the appearance of three major products, which can be separated. The glycan part of HGL consists of asparagine linked carbohydrates. HGL possesses four potential N-glycosylation sites (asparagine 15, 80, 252 and 308) with the consensus sequence Asn-X-Ser or Asn-X-Thr, but three of these were found to be occupied.

Several attempts at crystallizing purified native preparations of HGL and DGL were unsuccessful. The most successful crystals obtained from native DGL diffracted at a resolution of 4.5 Å. Charge homogeneity seemed to be essential to obtaining properly diffracting crystals. These crystals contained four to eight molecules per asymmetric unit in the case of DGL,. This pattern of cristallyzation in probably due to the large amounts of carbohydrates present in the protein.

BIOCHEMICAL ASPECTS

Acidic lipases are specifically inhibited by micellar diethyl p-nitrophenyl phosphate (E600), which is a well-known serine esterase inhibitor. Acidic lipase show the consensus sequence "(LIV)-X-(LIVFY)-(LIVST)-G-(HYWV)-S-X-G-(GZTAC)" located around serine 153, characteristic of serine enzymes. In pancreatic lipases, the active serine residue (Ser 152) is located at the hinge of the nucleophilic elbow: this residue has been found in other lipases to participate in the charge relay system, together with a histidine and an aspartic (or glutamic) residue. Tetrahydrolipstatin, a chemical substance derived from lipstatin produced by *Streptomyces toxitricini*, inhibited pancreatic and gastric lipases *in vitro* when emulsified tributyrin or a dicaprin monolayer were used as substrates. This inhibitor is linked to serine 152 in the case of pancreatic lipase.

Three cysteine residues are present in both HGL and DGL. Titration of HGL and DGL was performed by incubating the lipase with classical sulfhydryl reagents such as dithio-nitrobenzoic acid (DTNB), lauroyl thio-nitrobenzoic acid (C12-TNB) or di (thiopyridine) (4-PDS), and one titrable sulfhydryl group was found to exist per molecule of enzyme. Further addition of a denaturing agent such as SDS or urea did not increase the number of modified SH groups. As only the free sulfhydryl group was titrated, it can be concluded that gastric lipases contain a single disulphide bridge. The modification of the sulfhydryl group was accompanied by a concomitant loss of gastric lipase activity. It can be concluded that HGL and DGL possess a sulfhydryl group that seems to be either directly or indirectly involved in their catalytic activity. Using the monomolecular film technique, it was established that after the modification of one sulfhydryl group, gastric lipase still binds to films of phosphatidylcholine or dicaprin. The modification of the sulfhydryl group therefore seems to prevent the access of the substrate to the catalytic site and not to interfere with the interfacial binding step.

EXPRESSION OF GASTRIC LIPASES

Expression of Gastric Lipases in *E. coli*

In a preliminary study, HGL cDNA was fused with the maltose binding protein cDNA and expressed in *E. coli*. A soluble fusion protein of about 80 kDa was produced, which was recognized by rabbit polyclonal antibodies, but showed no activity towards triglycerides. Other attempts to produce soluble HGL were made, using pET plasmids such as pET 22b (+) plasmid. This vector carries an N-terminal pel B signal sequence for potential periplasmic localization. It also contains the natural promoter and coding sequence of the *lac* repressor (*lac I*), oriented so that the T7 *lac* and *lacI* promoters diverge. This vector was used to express HGL in *E coli* strains: BL21 (DE3) and BL21 (DE3) pLysS, which is lysogenic for λDE3, and carries the T7 RNA polymerase gene under *lac UV5* control. In the absence of IPTG, the *lac* repressor acts both at the *lac UV5* promoter by repressing the transcription of the T7 RNA polymerase gene by the host polymerase and at the T7 *lac* promoter, by blocking the transcription of the target gene by any T7 RNA polymerase which occurs. HGL expression in this system was modulated by modifying IPTG addition as well as the *E. coli* cell culture temperature during the induction phase. When *E. coli* cells were grown at 37°C, HGL was mostly produced in the form of insoluble material (inclusion bodies), whereas when the induction was performed at 15 or 20°C, a soluble protein was produced, that was recognized by anti-HGL antibodies with an apparent molecular mass of 43 kDa on SDS/PAGE gels. In all the cases tested, however, using the pH-state technique, the protein was inactive towards tributyroylglycerol. Similar results have been obtained with the cDNA coding for DGL. These data seem to indicate that the expression of a catalytically active lipase requires glycosylation. Expression experiments were therefore performed with HGL in eukaryotic systems, with which post transcriptional modifications of the enzyme such as glycosylation and disulphide bond formation can be performed.

Expression of Human Gastric Lipase in Insect Cells

The baculovirus/insect cell system

Baculoviruses belong to a group of diverse large double stranded DNA viruses that infect many different species of insects and arthropods, which are their natural hosts. Baculovirus strains are highly species-specific and are not known to infect any vertebrate hosts. The most extensively studied baculovirus strain so far is the *Autographa californica* nuclear polyhedrosis virus (AcNPV). The baculovirus infection cycle can be divided in two phases: the early phase in which extracellular virus particles

(ECV) sprout from the cell membrane of the infected insect cells, and the later phase, during which occlude virus particles (OV) are assembled inside the nucleus. The protein polyhedrin is the main component of the occlussion bodies. This protein, which is

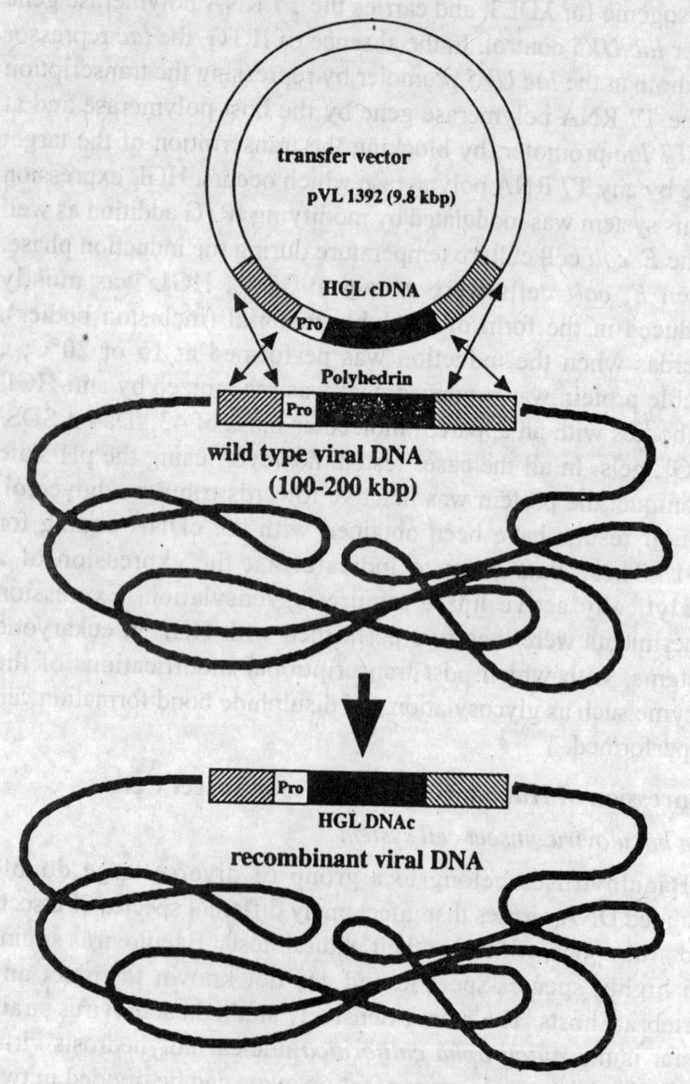

Figure 12.1: *Construction of HGL recombinant baculovirus*

expressed in large amounts, is not essential to the baculovirus life cycle and its synthesis is directed by a strong promoter. It can therefore be replaced by another gene, such as a gastric lipase gene, and expressed in insect cells. The gene of interest is transferred to the baculovirus by means of a transfer vector which contains and *E. coli* origin of replication as well as an ampicillin resistance gene and identical sequence stretches to those flanking the polyhedrin gene in the wild type genome. During cotransfection of insect cells by the baculovirus and the transferred gene, homologous recombination between these sequences occurs, resulting in a recombinant virus containing the gene of interest (Figure 12.1).

Expression of recombinant HGL (r-HGL)

Construction and expression of the recombinant baculovirus in insect cells

The gene from HGL was cloned into the baculovirus transfer vector PVL 1392 downstream of the strong polyhedrin promoter. The recombinant transfer vector was cotransfected with the DNA baculovirus, BaculoGold virus from pharmingen. BaculoGold™ DNA contains a lethal deletion and does not code for viable viruses by itself, but cotransfection with the recombinant transfer vector restores the deletion and yields a viable virus with a recombination efficiency of practically 100%. The recombinant virus stock was amplified by performing several rounds of growth and a high titer virus stock solution was harvested. For the expression of recombinant protein, BTI-TN-5B1-4 (or High Five™) insect cells were used, which were grown in a serum-free Ex-cell™ 400 medium.

Production and purification of r-HGL in tissue culture plates

The insect cells were infected with a multiplicity of infection (MOI) of between 1 and 5. The amount of HGL accumulated in the supernatant of the insect cells was monitored directly at various post-infection times by measuring the lipase activity.

The highest HGL activity, 8 units/ml of cell culture, was recorded 3 days post infection. r-HGL was secreted in an almost pure form in the supernatant of insect cell culture media during the first three days post-infection. From day 3 onwards, the activity decrease slightly with time. This decrease was consistent with the data available on cell lysis and the proteolytic degradation of r-HGL. The release of insect cell lysosomal proteases from lysed cells may attributable to the degradation of r-HGL in the culture medium.

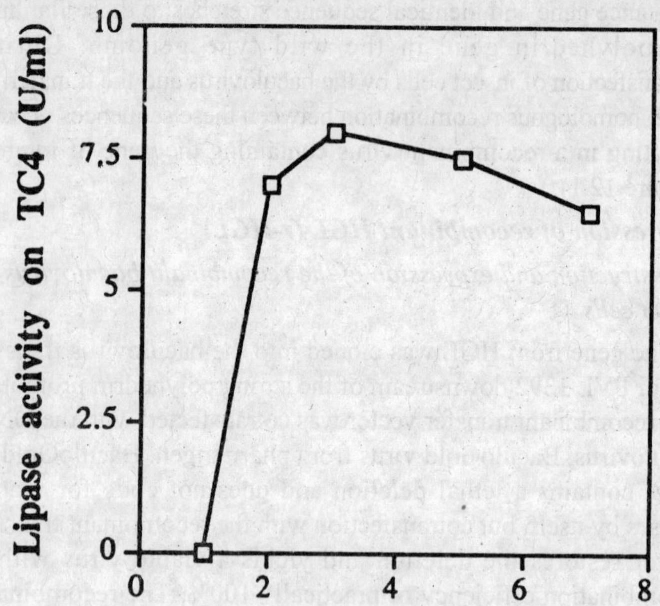

Figure 12.2: ***Kinetics of the onset of r-HGL activity in the supernatant of insect cells growin in monolayer flasks infected with HGL recombinant baculovirus. Monolayers of BTI-TN-5B 1- 4 cells were grown in serum free medium (Ex-Cell™ 400) in 75-ml flasks prior to infection at a MOI = 3.1 ml of supernatant corresponds to 5 × 10^6 cells***

r-HGL was purified in a single step by performing cation-exchange chromatography on SP-Sepharose.

Production and purification of r-HGL in bioreactor

Insect cells can be grown in large-volume suspension bioreactors. The possibility of producing the recombinant protein in a bioreactor is a great improvement over the use of flasks with monolayer cell cultures, since the handling of the insect cells is facilitated and large amounts of recombinant protein can be obtained in this way. The number of insect cells increases two-fold every 18–24 hours in shaker cultures, as compared with the production time of 28–30 hours in the case of monolayer cultures. Agitation can easily damage insect cells however, and the productivity of the system depends greatly on a number of factors, including the cell line, the age and condition of the cells at time of infection and the number of passages of the virus inoculum. We have used High Five™ cells, which were reported to be an excellent host for the expression of recombinant proteins, and several cell infection procedures with the MOI and the host cell density at the time of infection as variables. As the post-infection addition of glucose can facilitate enhanced recombinant protein production, we kept glucose concentration at 3 mg/ml by adding a concentrated solution every two-days. Two procedures are described below, the one based on a low MOI and a low cell mass, with which all the viral particles become attached to the cells (which are in great excess), and the other based on a high cell mass and a high MOI.

Insect cells obtained from *Trichoplusia ni*, BTI-TN-5BI-4 were grown in suspension and in serum free medium Ex-Cell™ 400. An inoculum was produced in spinner flasks (INTEGRA Biosciences) by infecting 600 ml of host cells (2.5 10^6 cells per ml) with a low viral titer solution (MOI = 0.1) of the recombinant baculovirus, and incubated for 3 hrs before being transferred to a 4-litre working volume bioreactor (Cytoflow, INCELTECH-France). The cells were then diluted down to a density of 6 10^5 cells per ml with fresh medium. In order to prevent the cells from aggregating, 0.1% (v.v) pluronic F-68 acid (from Sigma) was added to the medium. A constant glucose concentration of 3 mg/m. was maintained (Figure 12.3a). The r-HGL production was monitored on line by measuring the enzyme activity in 3-ml samples collected from the bulk solution. The cell density and cell viability were determined using the tryptan blue exclusion method with a Malassez haemocytometer. Centrifugation was performed at 5000 rpm for 10 min and cell supernatant was collected when cellular lysis reached 60–70%. With the second procedure, a large cell mass was obtained after growing the host cells in the 4-litre bioreactor for four days until a cell denstity of 3 10^6 cells/m. was reached and then infecting the cells with a high viral titer solution (MOI = 5,.

The recombinant protein yield differs considerably between those two procedures, since the method which consists of using a low cell density and a low MOI is about 4 times more efficient than that involving a large cell mass and a high MOI is about 4 times more efficient than that involving a large cell mass and a high MOI. Under the low conditions, about 8–20 mg of recombinant HGL (depending on the runs) were harvested in 1.5 1 culture medium on day 6–7 post infection. The two procedures actually differ not only the initial cell mass and the yield, but also in the dynamic interactions occurring among the propagating viruses, the cell division processes, the availability of essential nutrients, and the accumulation of waste metabolites. With the low procedure, the cell population is doubled at least once after the infection (infected cells do not multiply). The high productivity obtained under these conditions was directly due to the cells being infected in a relatively early phase of growth (*i.e.* in an exponential phase). In addition, with this method, the virus replication and subsequent horizontal spread of the infection take place in a better nutritional environment. The low procedure is the more efficient of the two, probably because in this case the recombinant protein expression is completed before the depletion of nutrients, as observed by other authors. On the other hand, with the high MOI-high cell mass protocol, the infection is performed into an already worked-out medium (three to four days of culture at high cell density), a rapid arrest of cell growth occurs and the cell viability decreases. Similar results have been obtained elsewhere by Yang *et al*, (1996) : these authors developed a serum-free cell culture process using a recombinant baculovirus to infect *Trichloplusia ni* insect cells to produce human lysosomal glucocerebrosidase. They developed a model for predicting the recombinant protein yield based on the key assumption that maximum protein production was limited by the nutritional value of the medium. Based on this model, it was predicted that optimum protein expression would be achieved with a 4-day batch process at a low

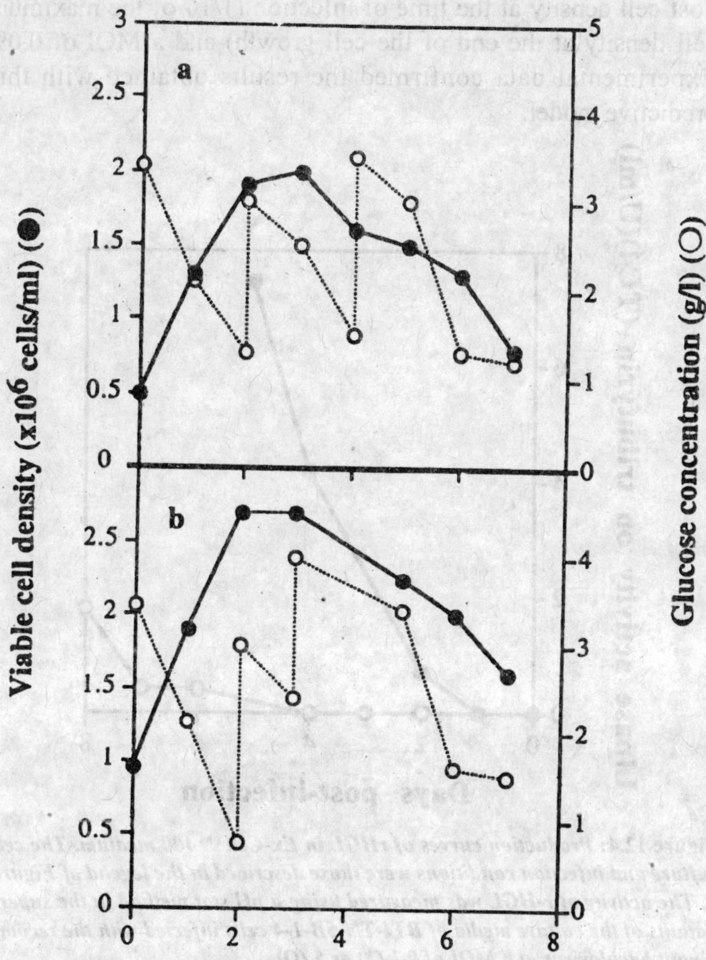

Figure 12.3: ***Growth curve of BTI-TN-5B 1-4 insect cells grown in Ex-Cell™ 400 medium using a 4-litre bioreactor, (a) The cell culture (2.2 × 10⁶ cells/ m.) was infected at a MOI of 0.1 in a spinner flask for 3 hours, transferred to a bioreactor and diluted. The cell concentration and residual glucose concentration were quantified once a day during the bioreactor run. Glucose was added to the bioreactor culture mdium on post-infection days 2 and 4. (b) The cell culture (3 × 10⁶ cells/ml) was infected at a MOI of 5 in a bioreactor. The cell concentration and residual glucose concentration were quantified once a day during the bioreactor rn. Glucose was added to the bioreactor culture medium on post-infection days 2 and 3***

host cell density at the time of infection (1/10 of the maximum cell density at the end of the cell growth) and a MOI of 0.09. Experimental data confirmed the results obtained with this predictive model.

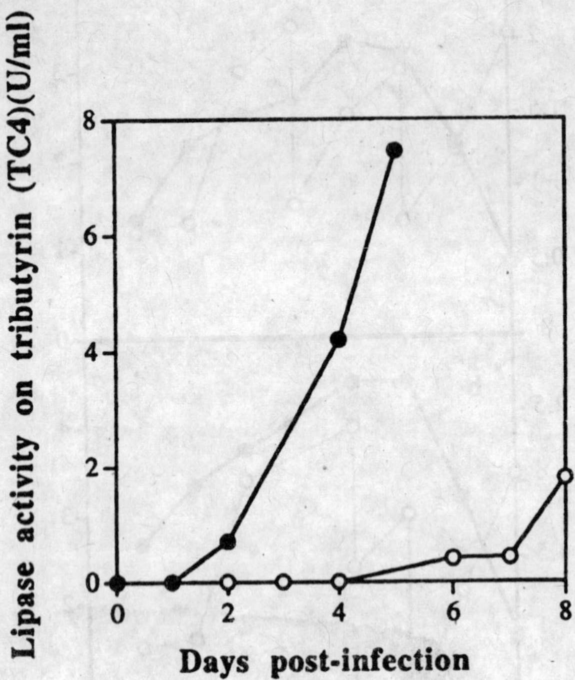

Figure 12.4: ***Production curves of rHGL in Ex-Cell™ 400 medium. The cell culture and infection conditions were those described in the legend of Figure 3, The activity of r-HGL was measured using a pH stat method in the supernatants of the culture media of BTI-TN-5B-1-4 cells infected with the recombinant baculovirus at a MOI of 0.1 (*) or 5 (O)***

One of the drawbacks of using of suspension cultures for this purpose is that a high rate of cell lysis occurs and intracellular proteins are discharged from the insect cells into the culture medium. During the r-HGL purification procedure, some problems arose due to the strong tendency of these lipases to adsorb on hydrophobic surfaces. The decrease in the HGL glycosylation level in the baculovirus/insect cells system may account for the decrease in solubility of the enzyme. We observed in some cases that the specific activity of r-HGL decreased in the culture medium after post-infection day 6. This result suggests that r-HGL decreased in the culture medium after post-infection day 6. This result suggests that r-HGL might aggregate, since no degradation fragments were detected in the immunoblotting experiments.

We nevertheless succeeded in purifying r-HGL by performing two chromatographic steps (SP-Sepharose followed by an anti HGL immunoaffinity column, data not shown).

Biochemical characterization of r-HGL

Recombinant HGL was identified using the Western blotting procedure with rabbit polyclonal antibodies. The fact that the protein migrated with an apparent molecular mass of 45 kDa under SDS-PAGE analysis (compared with 50 kDa in the case of natural HGL) indicates that insect cells have only a limited HGL glycosylation capacity.

The amino acid sequence of r-HGL has been found to resemble that of native HGL. The N-terminal protein sequence (LFGKL.......) determined from the r-HGL secreted by insect cells infected wit the recombinant baculovirus containing its natural signal peptide nucleotide sequence indicates that the cleavage site occurs at the expected position, between Gly -1 and Leu +1, as observed with native HGL. The amino acid sequence of r-HGL was determined up to residue 16. Asn 15 was detected, indicating that this residuce is probably not glycosylated. We are unable to detect any residue at position 15 of native HGL, however, which suggests the presence of a glycan chain. Lamb pregastric lipase has also been found to be glycosylated at that position. This result confirms that heterologous recombinant proteins are less glycosylated by insect cells than by mammal cells.

The pI patten of r-HGL displays three isoforms, with calculated pI of 7.0 (minor band), 7.5 and 8.1. Some heterogeneity therefore existed as the result of glycosylation, although it was less marked than in the natural form.

r-HGL was assayed on short-, medium-and long-chain triglycerides. The maximum specific activities of the r-HGL were 434 (Intralipid™), 730 (trioctanoylglycerol) and 562 (tributyroylglycerol) units/mg.

POTENTIAL APPLICATIONS

Substitutive Enzyme Therapy

Exocrine pancreatic insufficiency, which occurs, for example, in cystic fibrosis (1/2500 children) always gives rise to two serious problems: malnutrition and steatorrhea. These problems can be partly solved by the administration of porcine pancreatic exracts as a replacement therapy. To reduce the malabsorption of fat, the enzymes delivered into the duodenum must amount to 5–10% of the quantities usually present after maximum stimulation of the pancreas. Under optimum conditions, if no inactivation of the supplement enzymes occurs in the stomach or duodenum, approximately 30,000 IU of lipase must be taken with each meal, which corresponds to about 5–10 g of lyophilised pancreatic powder per day. In the past, preparations of this kind were far from satisfactory, since a large proportion of the enzymes administered were denatured in the stomach due to the extreme acidity of the gastric juice. Pancreatic lipase is irreversibly inactivated at pH levels downward of 4.0. This problem can be partly solved by using enteric-coated microspheric pancreatic preparations. The lipase activity is released from the microspheres at pH levels of around 5.5. It is necessary to bear in mind, however, that a considerable loss of activity occurs along the gastrointestinal tract, since only 8% of the exogeneous lipase was recovered in the above study. The lipase was degraded at the duodenal level by proteases present in the pancreatic extracts.

The use is acidic-resistant lipases should in principle yield more satisfactory results than the pancreatic preparations currently in use. The co-administration of acidic lipases, which hydrolyse dietary lipids under acidic conditions, should help to treat patients with various forms of pancreatic deficiency. Physiological studies have shown that preduodenal lipases are capable of acting not only in the stomach but also in the duodenum in synergy with a pancreatic lipase. Various clinical studies have been conducted on both animals and humans to assess the efficacy of enzymatic replacement therapies using acid-resistant lipases of bacterial origin to treat exocrine pancreatic insufficiency. Although this treatment significantly increased the weight and reduced the steatorrhea in dogs, the use of these enzymes is limited because of their sensitivity to the proteolytic action of gastric pepsin, as well as by legal requirements.

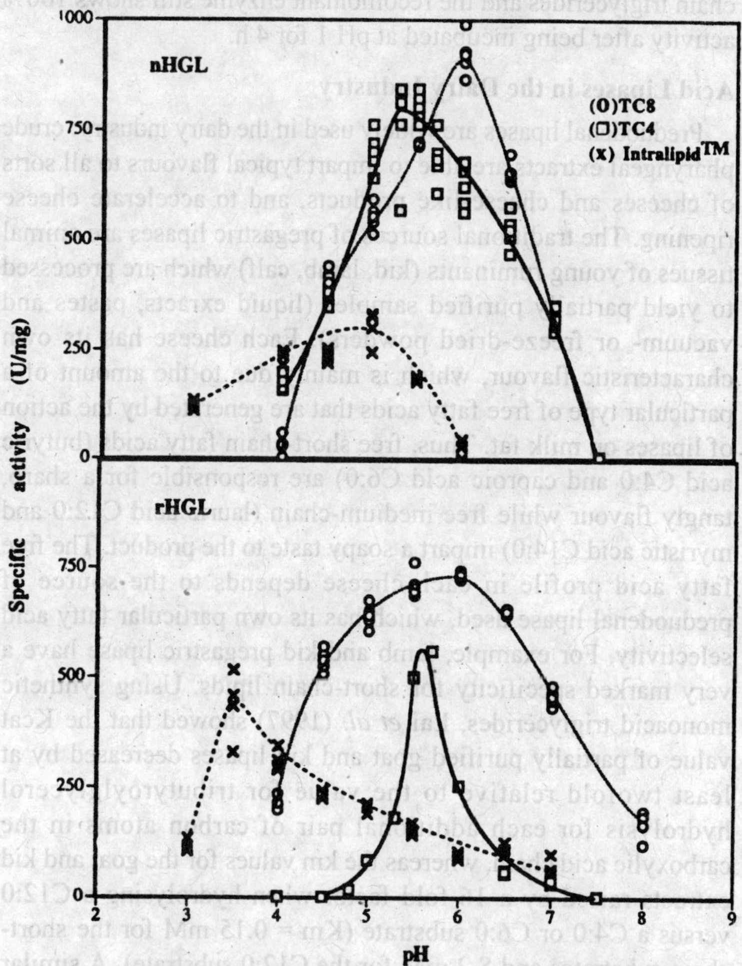

Figure 12.5: ***Effects of pH on the specific activity of natural HGL (upper panel) and r-HGL (lower panel) on tributyroylglycerol (TC4), trioctanoylglycerol (TC8) and long-chain (Intralipid™) triacylglycerols***

Unlike acidic lipases of bacterial origin, gastric lipases are resistant to digestive enzymes. Furhermore, they are active on long-chain triacylglycerols, which account for most of the dietary fat in humans (specific activity of HGL and DGL in Intralipid™: 600 U/mg and 950 U.mg, respectively). r-HGL displays interesting properties, since its pH optimum was found to be 3.5 on long-chain triglycerides and the recombinant enzyme still shows 100% activity after being incubated at pH 1 for 4 h.

Acid Lipases in the Dairy Industry

Preduodenal lipases are widely used in the dairy industry: crude pharyngeal extracts are able to impart typical flavours to all sorts of cheeses and cheese-like products, and to accelerate cheese ripening. The traditional sources of pregastric lipases are animal tissues of young ruminants (kid, lamb, calf) which are processed to yield partially purified samples (liquid exracts, pastes and vacuum- or freeze-dried powders). Each cheese has its own characteristic flavour, which is mainly due to the amount of a particular type of free fatty acids that are generated by the action of lipases on milk fat. Thus, free short-chain fatty acids (butyric acid C4:0 and caproic acid C6:0) are responsible for a sharp, tangly flavour while free medium-chain (lauric acid C12:0 and myristic acid C14:0) impart a soapy taste to the product. The free fatty acid profile in each cheese depends to the source of preduodenal lipase used, which has its own particular fatty acid selectivity. For example, lamb and kid pregastric lipase have a very marked specificity for short-chain lipids. Using synthetic monoacid triglycerides, Lai *et al.* (1997) showed that the Kcat value of partially purified goat and kid lipases decreased by at least twofold relative to the value for tributyroylglycerol hydrolysis for each additional pair of carbon atoms in the carboxylic acid chain, whereas the km values for the goat and kid extracts raised by a 16-fold factor when hydrolysing a C12:0 versus a C4:0 or C6:0 substrate (Km = 0.15 mM for the short-chain substrates and 8.2 mM for the C12:0 substrate). A similar reactivity with natural milk fat triglycerides was also observed. Thus, each preduodenal lipase impart its own characteristic flavour profile: a buttery and slightly peppery flavour (calf), a sharp "piccante" flavour (kid), a strong "pecorino" flavour (lamb). Kid or lamb pregastric extracts are used in manufacturing Romano, Domiati and Feta cheeses, calf or kid pre gastric lipases for Mozarella, Parmesan and Povolone cheese while a combination of calf gastric lipase and goat pregastric lipase can produce Cheddar or Provolone cheese. Cheese flavour is also function, among other factors, on the composition of milk fat: for example, ovine milk is richer in short-and medium- chain triglycerides, bovine milk fat is richer in long-chain and unsaturated triglycerides, and caprine milk fat is the richest in polyunsaturated medium-chain triglycerides. It is nevertheless possible to use preduodenal lipases for imitation of cheese made from ewe's or goat's milk. Addition of pre-gastric lipase from goat or kid to bovine milk fat leds to free fatty acid profile for hydrolysis of the bovine milk fat by goat and kid extract very similar to the one of Parmesan cheese. It should be noticed that short-chain fatty acids are available in the highest proportions at the sn-3 position of bovine milk fat (28.4% C4:0). Goat and kid pregastric lipases selectively hydrolyze triglycerides at this position. Positional selectivity alone cannot, however, account for the high concentration of butyric acid obseved after hydrolysis by goat and kid lipases (40 and 42.5%, respectively). One can imagine that recombinant technologies will made available in the future numerous acid lipases with finely tuned fatty acid specificities.

Index

A

B

C

D

E